Vol. 1
Computer Techniques

Computer Aided and Integrated Manufacturing Systems

A 5-Volume Set

Vol.1
Computer Techniques

Computer Aided and Integrated Manufacturing Systems

A 5-Volume Set

Cornelius T Leondes
University of California, Los Angeles, USA

World Scientific
New Jersey • London • Singapore • Hong Kong

Published by

World Scientific Publishing Co. Pte. Ltd.

5 Toh Tuck Link, Singapore 596224

USA office: Suite 202, 1060 Main Street, River Edge, NJ 07661

UK office: 57 Shelton Street, Covent Garden, London WC2H 9HE

British Library Cataloguing-in-Publication Data
A catalogue record for this book is available from the British Library.

COMPUTER AIDED AND INTEGRATED MANUFACTURING SYSTEMS
A 5-Volume Set
Volume 1: Computer Techniques

ISBN 981-238-339-5 (Set)
ISBN 981-238-983-0 (Vol. 1)

Typeset by Stallion Press

Printed by FuIsland Offset Printing (S) Pte Ltd, Singapore

Computer Technology

This 5 volume MRW (Major Reference Work) is entitled "Computer Aided and Integrated Manufacturing Systems". A brief summary description of each of the 5 volumes will be noted in their respective PREFACES. An MRW is normally on a broad subject of major importance on the international scene. Because of the breadth of a major subject area, an MRW will normally consist of an integrated set of distinctly titled and well-integrated volumes each of which occupies a major role in the broad subject of the MRW. MRWs are normally required when a given major subject cannot be adequately treated in a single volume or, for that matter, by a single author or coauthors.

Normally, the individual chapter authors for the respective volumes of an MRW will be among the leading contributors on the international scene in the subject area of their chapter. The great breadth and significance of the subject of this MRW evidently calls for treatment by means of an MRW.

As will be noted later in this preface, the technology and techniques utilized in the methods of computer aided and integrated manufacturing systems have produced and will, no doubt, continue to produce significant annual improvement in productivity — the goods and services produced from each hour of work. In addition, as will be noted later in this preface, the positive economic implications of constant annual improvements in productivity have very positive implications for national economies as, in fact, might be expected.

Before getting into these matters, it is perhaps interesting to briefly touch on Moore's Law for integrated circuits because, while Moore's Law is in an entirely different area, some significant and somewhat interesting parallels can be seen. In 1965, Gordon Moore, cofounder of INTEL made the observation that the number of transistors per square inch on integrated circuits could be expected to double every year for the foreseeable future. In subsequent years, the pace slowed down a bit, but density has doubled approximately every 18 months, and this is the current definition of Moore's Law. Currently, experts, including Moore himself, expect Moore's Law to hold for at least another decade and a half. This is hugely impressive with many significant implications in technology and economics on the international scene. With these observations in mind, we now turn our attention to the greatly significant and broad subject area of this MRW.

"The Magic Elixir of Productivity" is the title of a significant editorial which appeared in the *Wall Street Journal*. While the focus in this editorial was on productivity trends in the United States and the significant positive implications for the economy in the United States, the issues addressed apply, in general, to developed economies on the international scene.

Economists split productivity growth into two components: Capital Deepening which refers to expenditures in capital equipment, particularly IT (Information Technology) equipment: and what is called Multifactor Productivity Growth, in which existing resources of capital and labor are utilized more effectively. It is observed by economists that Multifactor Productivity Growth is a better gauge of true productivity. In fact, computer aided and integrated manufacturing systems are, in essence, Multifactor Productivity Growth in the hugely important manufacturing sector of global economics. Finally, in the United States, although there are various estimates by economists on what the annual growth in productivity might be, Chairman of the Federal Reserve Board, Alan Greenspan — the one economist whose opinions actually count, remains an optimist that actual annual productivity gains can be expected to be close to 3% for the next 5 to 10 years. Further, the Treasure Secretary in the President's Cabinet is of the view that the potential for productivity gains in the US economy is higher than we realize. He observes that the penetration of good ideas suggests that we are still at the 20 to 30% level of what is possible.

The economic implications of significant annual growth in productivity are huge. A half-percentage point rise in annual productivity adds $1.2 trillion to the federal budget revenues over a period of 10 years. This means, of course, that an annual growth rate of 2.5 to 3% in productivity over 10 years would generate anywhere from $6 to $7 trillion in federal budget revenues over that time period and, of course, that is hugely significant. Further, the faster productivity rises, the faster wages climb. That is obviously good for workers, but it also means more taxes flowing into social security. This, of course, strengthens the social security program. Further, the annual productivity growth rate is a significant factor in controlling the growth rate of inflation. This continuing annual growth in productivity can be compared with Moore's Law, both with huge implications for the economy.

The respective volumes of this MRW "Computer Aided and Integrated Manufacturing Systems" are entitled:

Volume 1: Computer Techniques
Volume 2: Intelligent Systems Technology
Volume 3: Optimization Methods
Volume 4: Computer Aided Design/Computer Aided Manufacturing (CAD/CAM)
Volume 5: Manufacturing Process

A description of the contents of each of the volumes is included in the PREFACE for that respective volume.

Computer Techniques is the subject for Volume 1. In this volume, computer techniques are shown to have significance in the design phase of products. These techniques also have implications in the rapid prototyping phase of products, automated workpiece classification, reduction or elimination of product errors in manufacturing systems, on-line process quality improvements, etc. These and numerous other topics are treated comprehensively in Volume 1.

As noted earlier, this MRW (Major Reference Work) on "Computer Aided and Integrated Manufacturing Systems" consists of 5 distinctly titled and well-integrated volumes. It is appropriate to mention that each of the volumes can be utilized individually. The significance and the potential pervasiveness of the very broad subject of this MRW certainly suggests the clear requirement of an MRW for a comprehensive treatment. All the contributors to this MRW are to be highly commended for their splendid contributions that will provide a significant and unique reference source for students, research workers, practitioners, computer scientists and others, as well as institutional libraries on the international scene for years to come.

Contents

CHAPTER 1

COMPUTER TECHNIQUES AND APPLICATIONS IN THE CONCEPTUAL DESIGN PHASE OF MECHANICAL PRODUCTS

WYNNE HSU and IRENE M. Y. WOON

School of Computing
National University of Singapore
Lower Kent Ridge Road
Singapore 119260
{whsu,iwoon}@comp.nus.edu.sg

The conceptual stage of the design process is characterized by a high degree of uncertainty concerning the design requirements, information and constraints. However, decisions made at this early stage have a significant influence on factors such as costs, performance, reliability, safety and environmental impact of a product. More importantly, a poorly conceived design can never be compensated for in the later stages of design. There is some controversy over the use of computers at this stage of product design. Some researchers feel that providing accuracy during this phase when solutions are imprecise, ill defined, approximate or unknown, accurate calculations impart a false sense of confidence in the validity of the solution. Others feel that maturing computer techniques with richer representations can provide invaluable assistance in specific sub-tasks of this phase. The purpose of this paper is to review advances in computational support for conceptual design from its early days to its current position. For each technique, we follow its progress from its conception to its latest status, pointing out significant variations and trends.

Keywords: Conceptual design; computer-aided design; mechanical product design; conceptual design models.

1. Introduction

The conceptual stage of the design process is one of the most imaginative stages of the design process in which human creativity, intuition and successful past experience play an important role. This early stage of the design is identified with a high degree of uncertainty concerning the design information and lack of clarity of the design brief (i.e. mission, instructions). The design of mechanical products is complex (as opposed to well-defined domains such as VLSI design) because they are, in general, multi-faceted. Some attributes of the task related to conceptual design process can be summarized as:

(1) Analysis of many dimensions of the problem in search of possible solutions.
(2) Synthesis of a number of possible solutions within a framework of constraints and requirements set forth in the design brief.
(3) Critical evaluation of alternative solutions.
(4) Selection of the design option that best fits the purpose.

These activities are highly non-linear and non-algorithmic by nature. There are no predefined rules for formulating design solutions. At present, design solutions developed mainly rely on heuristics and past experience.

There is some controversy over the use of computers in the early stages of product design. One school of thought is of the opinion that at the early stages of design where solutions are ill defined, accurate calculations impart a false sense of confidence in the validity of the solution. They support the practice of using heuristics that are relatively simple and less accurate than algorithmic techniques. The counter argument is that computers can generate and handle complex representations with ease and so even at the very early stages of the design process, one could introduce advanced algorithmic techniques. Hence, even though the designer may not have determined the design parameters to a high level of accuracy, one has not introduced further inaccuracy through the algorithm. In addition, there is also the concern that the instant we analyze a situation in terms of properties, artifacts, etc. we limit our view of the problem to that which can be expressed in modeling paradigm. For example, in the expert system to select a bridge type, only the structural designs built into the expert system can be designed. This creates 'blindness' for all other kinds of possible designs. The counter argument is that as computer techniques mature, with richer representations, more background knowledge and deep knowledge (or reasoning from first principles), they can perform well in real-world applications. This avoids the problem of blindness creation in the early stages of product design.

The purpose of this paper is to review advances in computational support for conceptual design from its early days to its current position. We define 'early days' as techniques and applications that originated before the 1980s, the 'recent past' as developments that originated from 1980s–1990s, and the 'current scene', as developments that germinated from 1997. For each technique, we follow its progress from its conception to its latest status, pointing out significant variations and trends. The later techniques tend to exhibit a 'hybrid' approach, reflecting the inadequacy of any single technique in supporting this complex phase of product creation.

2. Early Days

The predominant techniques used to support conceptual design in its early days (1960s to 1980s) are systems that were built on languages, images, graphs and operation research techniques. Computer technology, both hardware and software, were immature at this time. In this section, we look at how such systems have matured over the years to support the complex task of conceptual design.

2.1. *Languages*

Language represents an attempt at formalizing design. It is useful in expressing our understanding of designs unambiguously. In general, a language is defined by a grammar. A grammar is denoted by the quintuple (T, N, S, P) where T is the set of terminals, N is the set of non-terminals, S is the start symbol and P is the set of production rules. Table 1[1] gives an example of how part of a grammar (expressed in BNF specification language) can be used to describe the positions and motions of each part of a fixed axes mechanism and their relationship between them. The terminals are expressed in bold fonts. The non-terminals are expressed in normal fonts. The start symbol is *Motion* and the production rules are listed in Table 1.

Due to its compact representations, grammar/language is an efficient means of structuring design knowledge. Indeed, many pieces of work have used language/grammar as the underlying representation for their design knowledge. For example, Rinderle[2,3] used a graph-based language to describe behavioral specifications of design as well as the behavior of the components. Neville and Joskowicz[1] present a language for describing the behavior of fixed-axes mechanism e.g. couplers, indexers and dwells. Predicates and algebraic relations are used to describe the positions and motions of each part. Vescovi *et al.*[4] developed a language, CFRL, for specifying the causal functionality of engineered devices. In terms of grammar, Carlson,[5] Stiny[6] and Heisserman[7] have looked into using shape and/or spatial grammars to express physical design forms. In particular, Mitchell[8] has combined shape grammars with simulated annealing to tackle the problem of free-form structural design. First, shape grammars are used to generate structural design possibilities. Then, stochastic optimization of all the possible designs are achieved using simulated annealing. This allows the generation of large number of sound, efficient free-form solutions that otherwise would never have been imagined. A number of researchers[9–13] have also made use of grammars in engineering applications. Tyugu[14] also proposed an attribute model based on attribute grammar for representing implementation knowledge of design objects. Similarly, Andersson *et al.*[15] proposes a modeling language, CANDLE, which enables the use of engineering terminology to support early design phases of mechanisms and manipulator systems. In CANDLE, the basic taxonomies of engineering terminology are augmented with

Table 1. Example of part of a grammar.

Motion ::= SimpleMotion | ComplexMotion
SimpleMotion ::= <Part, SM_Type, Axis, InitialPosition, Extent, Relations>
SM_Type ::= **Translate** | **Rotate** | **Screw** | **Translateand Rotate** | **Stationary** | **Hold**
Extent ::= AxisParameter **by** Amount
Amount ::= **Real** | **Constant** | **Variable** | **Infinity**

the physical and solution principles that are specific for the design of mechanisms and manipulator systems.

In fact, the general approach adopted by researchers is that they would propose different special purpose languages to describe some aspects of design that they are interested in modeling. This approach of developing special-purpose languages works well for non-collaborative design effort. With the increasing emphasis on the use of product design as a firm's competitive advantage, the trend is towards supporting concurrent collaborative design. Hence, we find that effort is directed to developing shareable design ontology. An ontology is a useful set of terms/concepts that are general enough to describe different types of knowledge in different domains but specific enough to do justice to the particular nature of the task at hand. Alberts[16] proposed YMIR as an engineering design ontology. The "How Things Work" project at Stanford University[4] aims to build a large-scale ontology of engineering knowledge. By having a common set of ontology, knowledge can be reused and shared. This allows better integration between the different phases of the product's life cycle.[17]

2.2. *Graphs*

Graphs and trees are popular representations in the conceptual design stage. They have been used to model all aspects of a product — function, behavior, and structure. Function is the perceived use of the device by the human being. Behavior is the sequence of states in which the device goes through to achieve the function. Structure refers to the physical components or forms that are utilized to achieve the behavior. Kuipers[18] illustrates this distinction with the example of a steam valve in a boiler. The function of the steam valve is to prevent an explosion, its behavior is that it opens when a certain pressure difference is detected and its structure is the physical layout and connection between the various physical components.

Malmqvist[19] demonstrates how graphs can be used to model the functions of structural systems in mechanics, electronics, hydraulics e.g. hole punch, washing machine. Nodes of graphs are lumped elements which correspond to the different physical properties (capacitance, transformers) and these nodes are connected by edges (bonds) e.g. force, velocity. The power flow direction and causality of bonds are specified. Murthy and Addanki[20] manipulate a graph of models to modify a given prototype of some structural engineering system e.g. design of beams. A model describes the behavior of the system under certain explicit assumptions. The models form the nodes in a graph and the edges represent sets of assumptions that must be added or relaxed to go between adjacent models. Graph/trees have also been used to model the physical representations of the design components and their layout.[21,22] Besides modeling structural, behavioral and functional aspects of the product, graph and trees have also been used to model requirements and constraints.[23] Kusiak and Szczerbicki[24] use tree models in the specification stage of conceptual design to represent the functions and requirements of mechanical systems, with an incidence

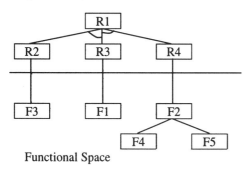

Requirement Space

R1: Design a shaft coupling
R2: Nature of the coupling is rigid
R3: Coupling is able to transmit torque
R4: Nature of the coupling is flexible

F1: Transmit energy
F2: Compensate offset of the shaft
F3: Connect two parts of the shaft rigidly
F4: Compensate offset applying a sliding element
F5: Compensate offset without applying a sliding element

Fig. 1.   An example of graph model.

matrix to represent the interaction between requirements and functions. Figure 1 shows the requirement and functional tree for the design of a shaft coupling.

An arc between the nodes of a tree represents a conjunction. A node without an arc represents a disjunction. There are therefore two sets of requirements that satisfy $R1$: $\{R2, R3\}$ and $\{R3, R4\}$.

2.3. *Images*

Perhaps the closest to human's way of thinking and reasoning is through the use of visual thinking models. Visual thinking has its beginning since 1969.[25] It did not gain a high profile in design research until McKim[26] demonstrated through experimental studies that visual thinking is vital to all branches of design practice. Freehand sketching is good for accelerating discussions and for comparing different solutions.[27]

Hand-sketched diagrams are also good in allowing different views of the sketch so as to obtain a good spatial image of the design solution.[28] In 1990, Radcliffe and Lee[29] proposed a model for the process of visual thinking that overcomes the barrier between the cognitive processes and the physical domain. Sittas[30] further explored the issues involved in supporting the creation and manipulation of 3D geometry during the conceptual design sketching activity.

2.4. *Operation Research models*

Operation Research (OR) emphasizes structured, numeric models, where a model is expressed in equations and the design goal, as one or many objective functions e.g. minimum weight, size or cost. Systems built with such underlying models endeavor to find values of variables that meet the equations and maximize/minimize one or several objective functions. In general, design problems are represented as follows. Let the continuous variables be x and the discrete variables be y.

The parameters which are normally specified as fixed values are represented by theta (θ). The design goal (or goals) can be expressed as the objective function $F(x, y, \theta)$. This function is a scalar for a single criterion optimization, and a vector of functions for a multi-objective optimization. Equations and inequality constraints can be represented as vectors of functions, h and g, that must satisfy,

$$h(x, y, \theta) = 0$$
$$g(x, y, \theta) \leq 0.$$

Many techniques have been proposed to solve optimization problem. A survey of the state-of-the-art optimization techniques in structural design can be found in Koski.[31] The focus of the survey is primarily based on the Pareto optimality concept. Briefly, Koski classified the multi-criteria structural design process into three phases. The first phase is the problem formulation where the criteria, constraints and design variables are chosen. The second phase is the generation of Pareto optimal solutions. The final phase describes the decision-making procedure employed to select the best compromise solution. In another paper by Levary,[32] he draws attention to the interaction between operation research techniques and engineering design. Specific applications of operation research methods are discussed with respect to the following engineering disciplines: computer engineering, communication system engineering, aerospace engineering, chemical engineering, structural engineering and electrical engineering.

A major advantage of OR models is that they provide a great deal of explanatory power in applications where they do apply. However, they are not always easy to apply because the data required by the algorithms may not be available, their scope of applicability is narrow and the algorithms used may not be able to provide optimal solutions because of the problem's complexity.

3. Recent Past

Conceptual design is an engineering activity that is generally ill-structured as it is performed early in a product life cycle, where complete and exact information and knowledge of requirements, constraints etc. is difficult to obtain. This highly skilled task is very complex and requires a mixture not only of different sources of knowledge (e.g. costing, performance, environmental issues) but also different types of knowledge (e.g. physical, mathematical, experiential).[33] The need to integrate different sources and types of knowledge is the emphasis of artificial intelligence research which gained prominence in the early 1980s and sparked off development in the areas such as object-oriented modeling, geometric modeling, case-based modeling and knowledge-based modeling. For example, object-oriented modeling has its roots in frames, an established knowledge representation scheme. Each of these areas has made significant impact on the conceptual design process, as we will see in the following subsections.

3.1. *Geometry models*

Geometry models focus on representing the structural aspects of a product. The objective is to represent 2-dimensional or 3-dimensional geometric shapes in a computer.[34] Popular representations of geometric shapes include: B-rep (boundary representation), CSG (constructive solid geometry), variational geometry and feature representations.

B-rep represents geometry in terms of its boundaries and topological relations. The transformation from one topology to another can be achieved using Euler operators. Since Euler operators are sound,[35] the topological validity of the structure is guaranteed. The major limitation of B-rep is its inefficiency in performing geometric reasoning. While in a B-rep approach, a shape is represented by the boundary information such as faces, edges and vertices, the CSG approach models geometric shapes using a set of primitives such as a cube, cylinder or a prism. Complex shapes are built from the primitives through a set of operators (union, difference and intersection). For example, the primitives given in Fig. 2 can be combined using set operations to form complex solids like that given in Fig. 3.

Although CSG is a geometry modeling technique that was widely accepted by both the research community and industry, it faces several inherent limitations. The most serious limitation, in our opinion, is the non-uniqueness of the CSG representations. This non-uniqueness of representations makes recognition of shapes from

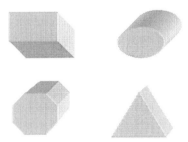

Fig. 2. Some CSG primitives.

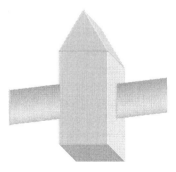

Fig. 3. Complex solid example.

CSG representation extremely difficult. Hence, this tends to dissuade researchers from relying solely on CSG representations alone. In addition, CSG representation does not guarantee that the solid it models is always a valid object. It is possible in CSG representation to model an invalid solid.

Variational modeling allows a designer to use equations to model mechanical components analytically and is popular because it allows the evaluation of competing alternatives. The concept of using variational geometry in computer aided design started as early as 1981. Lin[36] in his thesis described the feasibility of using variational geometry to model geometric information. Light and Gossard[37] expanded upon his work to allow modification of geometric models through variational geometry. Variational geometry design while general and flexible, necessitates the intensive use of numerical solvers to solve many simultaneous nonlinear equations. Frequently, solvers cannot solve these equations. Shpitalni and Lipson[38] combine parametric design with geometric design to ensure that the resulting system is both flexible and guaranteed to find a solution. This system was tested in the designing of sheet metal parts.

In the feature representation approach,[39,40] a part is built from a set of primitive building blocks with the guarantee that this set of building blocks are manufacturable. The notion of features was first proposed as form features[39,41] to bridge the gap between units of the designer's perception of forms and data in geometric models. Shapes are described as the way the designer understands them. A feature-based design approach allows a user to use mechanical features stored in a feature library in his design.[42]–[45] It provides a means for building a complete CAD database with mechanical features right from the start of the design. However, this approach suffers from the difficulty of a limited number of available feature primitives. It is difficult to satisfy various design needs and in the event that the features interact with one another, new features may arise that can cause complication in the analysis process. EDISON[46] is an example of a system using feature-based modeling. It has a database of known mechanisms and is indexed by their functions, structures and situations in which they are used. Thus far, the majority of feature-based research focuses on using feature-based design for process planning[47,48] and feature recognition.[49] Han and Requicha[50] proposed a novel feature finder that automatically generates a part interpretation in terms of machining features. The feature finder strives to produce a desirable interpretation of the part as quickly as possible. Alternate interpretations could be generated if the initial interpretation was found to be unacceptable by a process planner.

Recently, the trend has been towards the integration of various representation schemes. Keirouz et al.[51] proposed an integration of parametric, geometry, features, and variational modeling. With this integration, they showed that the system is able to handle geometry and "what if" questions arising in conceptual design.

In all the above approaches, the assumption is that the support of surface features is well defined on prismatic objects. This is not the case for sculptured surface models and current methods often lead to data explosion. Elsas and Vergeest[52] proposed a displacement feature modeling approach. In this approach, explicit modeling of protrusions and depressions is done in free-form B-spline surfaces that can achieve real-time response and with unprecedented flexibility.

3.2. *Knowledge-based models*

One major development in the recent past is the introduction of knowledge-based models. Knowledge-based models are used to capture procedural design knowledge as well as product or domain knowledge. A prominent branch of knowledge based models is the production model which uses rule representation to facilitate high-level reasoning. The rule based paradigm is adopted by Rao[53] to give advice on which alternative should be chosen in the design of ball bearings. An example of a rule is given in Fig. 4.

Besides rule representation, frame representation is also widely used. In the paper by Tong and Gomory,[54] he used a frame-based structure to model parts of standard kitchen appliances and light sources.

The underlying reasoning techniques used in production models include abductive, deductive, constraint-based, and non-monotonic reasoning. Abductive reasoning says that:

The surprising fact C is observed;
But if A were true, C would be a matter of course.
Hence there is reason to suspect that A is true.

In other words, abductive reasoning (goal directed) tries to derive the premises of a stated conclusion. On the other hand, deductive reasoning says that:

Suppose if A is true, then C would be a matter of course,
Now, we observe the fact A.
We can conclude that C is true.

Hence, deductive search (data driven) moves to arrive at some conclusion, given the initial facts.

An example of an abductive search strategy is given in Tong and Gromory[54] in the design of small electromechanical appliances. Rao[53] shows the use of deductive search strategy in selecting the appropriate ball bearings' design for a set of input parameters e.g. load type, bearing speed, environment of use, etc. Arpaia et al.[55] and Carstoiu et al.[56] makes use of both patterns of reasoning, the former in

If feature is SLOTA and
　　If interactingFeature is SlotB and
　　　　If typeofInteraction is intersecting then
　　　　　　Send the message intersectingWIth: SlotB to SlotA
　　　　　　To get edge entities of SlotA based on typeof Interaction

Fig. 4. An example of rule representation.

the design of measurement systems, in mapping from the logical attributes to the physical components of the instrument and the latter in the design of gears. Typically, abductive and deductive reasoning will face the problem of scaling-up. To address this problem, constraint-based reasoning is introduced. Further elaboration on constraint-based reasoning is given in Subsec. 3.3.

There are at present, a number of tools which couple knowledge based systems with conventional systems. Krause and Schlingheider[33] gives a comprehensive overview of such tools e.g. ICAD, MEDUSA-ENGIN, CONNEX. Increasingly, these tools are addressing the problematic areas of development and design.[57,58] Recent development has been towards the concept of metamodels. A metamodel is a qualitative model of causal relationships among all the concepts used for representing the design object.[59] The metamodel reflects the designer's mental model about the structure and behavior of the design object. Metamodel mechanisms include the primary model (a description of the requirement given by the designer) and aspect models (qualitative and quantitative models focusing on specific aspects of the design object).

Though there have been many successful applications that are built upon knowledge models presented here, a number of issues still remain unresolved. Some of these issues include: the verification of the correctness of knowledge models, the handling of incomplete knowledge, the resolution of inherent contradictions that are present in knowledge models and the incremental addition of new knowledge to existing knowledge models.

3.3. *Constraint-based models*

Constraint-based models rely on the designer's experience to select the bounds of design variables that define the search space in which the constraints are processed. A large search space as is expected in real world applications may have all the feasible solutions but it may contain a large number of infeasible solutions. However, if the search space is too restricted (as when particular variables have to be optimal), the risk is that no feasible solution exists in this small space.

A constraint is a statement about a design, the truthfulness of which does not depend on any tradeoffs with goals. For example, the manufacturing cost of the product of around $100 is a constraint whereas a manufacturing cost objective is to have the product manufactured at the lowest cost. In many instances, it may be possible to translate an objective into constraints e.g. the objective "minimize manufacturing cost" could be stated as manufacturing cost should be less than or equal to $100. Harmer *et al.*[60] shows how the functional requirements of a product is written as a set of constraints and translated into a desired property profile (which includes functions and objectives) to be matched against that of the existing components in an engineering catalogue. Kolb and Bailey[61] specify constraints between objects derived from analyzing the design of an aircraft engine, and employ a constraint propagation technique to integrate and perform mathematical analyses of the resulting solution which is the set of design parameters that satisfies all constraints. Oh *et al.*[62] give an example of how a constraint-based approach may result in the improved design of a video cassette tape.

In Ref. 63, constraint management moved away from emphasis on developing a strategy detection algorithm for designing bridges to a more human-centred approach where the designer is able to apply the heuristics they choose rather than a predefined set of heuristics. Vujosevic *et al.*[64] use a reason maintenance system to perform goal-directed search. An assumption-based truth maintenance system and multiple worlds are used to discover and store information about feasible designs and to avoid further consideration of infeasible design alternatives. Yao and Johnson[65] propose a domain propagation algorithm that is able to generate a more focused search space without omitting any feasible solutions in the original search space.

3.4. *Case-based models*

A consensus exists among AI researchers that reuse of the process of design rather than the product of the design might be more useful. In fact, much of design consists of re-design, in the adaptation of a previous design to a new context, or in the design iteration cycle. Case-based reasoning applies past experience stored in a computerized form towards solving problem in similar contexts. It involves three stages: the representation of cases, the matching and retrieval of similar cases, and the adaptation of the retrieved cases.

Case-based reasoning has been successfully applied where the structure and content of design information can be encoded symbolically and manipulated using artificial intelligence techniques. KRITIK[66] solves the function-structure design task in the domain of physical devices. Knowledge of previously encountered designs are organized as a design case which contains the functions it can deliver, and a pointer to the structure-behavior-function model for the design that explains how the structure of the design delivers its functions. The cases are indexed by the functions delivered by the stored designs. In CADET,[67] each case involves 4 different representations: object-attribute-value tuples, functional block diagrams, causal graphs and configuration spaces. Thus, all three levels of abstraction are represented and reasoning using the causal graph enables the structure-function transformation. If no case matches the current specification, transformations are applied to it until it resembles some case in the case database. Li *et al.*[68] employ a library of mechanical devices to aid in the design synthesis process. Gomes and Bento[69] proposed an algorithm for problem elaboration to ensure that problem specifications produced in early stages of design are complete and well defined. This algorithm used the functional, behavioral and structural knowledge stored in the cases and applies the knowledge to the layout design of bedrooms.

Sycara and Navinchandra[67] looks into the use of case representations to support conceptual design activities. In another work by Hsu *et al.*,[70] case representations have also been used to capture assembly-oriented design concepts. The case-based approach for storing feedback is rather natural and is a practical way to collect and store feedback that can be used for future projects. Information from the feedback is stored and linked directly to the part that is criticized. Irgens[71] has extended the scope of case library to include design intent, design data, and customer feedback, so as to provide a complete integrated historic advice for product prototyping. Simina and Kolodner[72] outlined a framework for creative design using case-base design and analogical understanding and reasoning. The system, ALEC allowed designers to explore the design space, encoding goals to recognized anomalies, ambiguities and issues. This process of exploration and tinkering allowed designers to recognize ideas for new projects. Other researchers like Mostow[73] and Banares-Alcantara *et al.*[74] are also experimenting with applying case-based reasoning to design plans.

Case-based reasoning techniques favor classes of domains where the number of primitive components are large as this ensures that the computational cost of retrieval and adaptation would be less than the cost of generating the solution from primitive components. On the other hand, case-based reasoning cases are stored over a long period of time and for that large number of cases, this may not be practical. To address the issue of a large number of cases, Murakami and Nakajima[75] proposed a computerized method of retrieving mechanism concepts from a library by specifying a required behavior using qualitative configuration space as a retrieval index. During retrieval, only mechanism concepts that realize specific kinematic behavior are retrieved. This effectively reduces the huge search space required.

3.5. *Objects*

An increasingly popular modeling representation is the object. An object is an entity that combines its data structure and its behavior into one. The advantages of object representation are abstraction (focus on what it does before deciding how to implement it), encapsulation (separating external aspects of an object which are accessible to other objects from the internal implementation details which are hidden from the other objects), polymorphism (do not consider how many implementations of given operation exist) and inheritance (of both data structure and behavior which allows sharing without redundancy). Figure 5 shows that a can-opener is a composite object made up of three other objects, with its corresponding object representation given in Fig. 6.

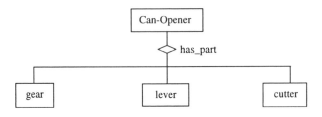

Fig. 5.   A can-opener object.

Instance	can-opener
Class	artifact
Initialization	
name	can-opener1
length	8
width	3
purpose	open cans
weight	10
material	steel
methods	(check_constraints(), assemble_part(), has_part())

Fig. 6.   The object representation of a can-opener.

Objects have been used to model many different kinds of entities. Martin and Roddis[76] proposed an object-oriented tree representation to model metal fatigue and fracture. In their object-oriented tree, each node is a "class". The root object represents the most general case of fatigue and fracture. Each class has a slot to represent the associated constraints and relationships. A similar approach has been taken by Ohki[77] to use object-oriented structure to represent constraints (law of physics) and physical objects (diode). In the domain of ship design, Yoshioka *et al.*[78] uses objects to represent the physical objects knowledge and the design process knowledge. Kolb and Bailey[61] use object-oriented techniques for modeling preliminary designs in the domain of aircraft engine design. Types of objects modeled include components (physical elements of a design), sub-models (properties of a design as a whole such as total weight, total cost), programs (external analysis codes for evaluating the design components), modules (simple design analyses), links (specifying constraints between objects). A novel approach to geometric reasoning using object-oriented approach was proposed by Nacaneethakrishnan *et al.*[79] whereby geometry is abstracted in terms of form features and the spatial relationships between features are represented using intermediate geometry language (IGL). Object algebra is then used to perform geometric reasoning. Kusiak *et al.*[80] use a hybrid of object-oriented representation and production rules in his CONDES system. The object-oriented representation is used to model design synthesis while the production rules are used to guide the process. Bento *et al.*[81] present a hybrid framework to integrate first-order logic into the object-oriented paradigm for representing engineering design knowledge. The logic component will be used for representing knowledge that is expected to be subjected to frequent changes throughout the design process, while objects are used to describe other pieces of knowledge whose structure is less likely to change.

3.6. *Qualitative models*

Qualitative Reasoning (QR) is defined as the identification of feasible design spaces using symbols and intervals of continuous variables. This allows formal simplified representations about a domain that maintains enough resolution to distinguish and explain the important features of behavior while leaving out the irrelevant details. Such representations are known as Qualitative Models. For example, we are interested in whether water in the pan is hot or cold rather than its exact numerical value. Qualitative Reasoning is therefore particularly pertinent in early design phases when little quantitative information is available. The device-centered ontology proposed by De Kleer and Brown[82] deals with the problem of deriving function or behavior of system given its structural descriptions and some initial conditions. All possible behaviors are determined by generate-and-test or constraint-satisfaction technique. Other researchers[83,84] have applied the process-centered ontology[85] to represent and reason about the states and behaviors of mechanical devices that handles kinematics and dynamics of mechanical devices.

EDISON[86] is a project that aims to construct an engineering design invention system by employing qualitative reasoning in the domain of mechanics. Work on this project has considered how functional knowledge can be integrated with qualitative reasoning. Many other researchers[87–89] have proposed the use of qualitative reasoning to structural engineering design problems. Fruchter[90] applies qualitative reasoning at different structural abstraction (structural, process and structure parameter) levels to select design modifications arising from performance problems of lateral load resisting frame structures. Murthy and Addanki[20] uses a graph of models approach in PROMPT to analyze a prototype based on first principles and derive its behavior from its structure. Li et al.[68] recently propose a combination of qualitative and heuristic approach to the conceptual design of mechanisms. The basic idea is to represent and classify a library of mechanical devices qualitatively and then employ best-first heuristic searches to generate a set of feasible design alternatives from a given specification.

One of the limitations of qualitative reasoning is the generation of spurious behaviors that is a result of ambiguity. However, De Kleer and Brown[82] see ambiguities as 'strong points'; they use it as a means to explore alternative interpretation of the same system. Ways of resolving ambiguity include maintaining information about partial ordering relations of parameters,[91] incorporating heuristics[92,93] and maintaining a semi-qualitative model.[94,95]

There is not enough resolution in qualitative representations to reason effectively about space, shape and spatial events. The ability to address this shortcoming will have important implications on the field of computer aided design. Some researchers working in this field are Gelsey,[96] Faltings[83] and Forbus et al.[97] Law[98] combines symbolic qualitative reasoning with diagrammatic reasoning to support the preliminary design of building structures. The symbolic reasoning component contains the symbolic representation of the structural components, qualitative structural engineering knowledge about joints, supports and other structural components and the constraints that they impose on the structure. The diagrammatic reasoning component includes an internal representation of the frame structure as well as a set of operators to manipulate and inspect the shape. This enables the system not only to reason about the problem and the domain, but also allows the representation of the kinds of operation that humans visually perform with a picture.

3.7. *Neural networks*

Artificial neural net models are interconnected neurons, each with some kind of internal activation function that allows the network to mimic the activities of the human brain to some extent.[99] Biological neurons transmit signals over neural pathways. Each neuron received signals from other neurons through special connections called synapses. Some inputs tend to excite the neuron while others tend to inhibit it. When the cumulative effect exceeds a threshold, the neuron fires and a signal is sent to other neurons. An artificial neuron receives a set of inputs. Each input is multiplied by a weight analogous to a synaptic strength. The sum of all the weighted inputs determines the degree of firing called the activation level. The input signal is further processed by an activation function to produce the output signal, which is transmitted along. A neural network is represented by a set of nodes and arrows. A node corresponds to a neuron, and an arrow corresponds to a connection between neurons. In general, neural networks are good for classification tasks and for performing associative memory retrieval. As a result, many neural networks applications in engineering design is geared towards either classifying the designs into families of design problems[100] or to find the nearest values for the design parameters.[101]

A common limitation of neural networks and genetic algorithms is that the design must be specified by a limited list of design attributes, which implies that the reasoning carried out is superficial based solely on the similarity of the attributes and their values. Selection of the number of training patterns is an extremely important issue in the performance of the network.[102] During the training process, the network learns from the experience and examples presented to it. If inadequate data is provided, generalization will be difficult and the network response to unknown data, poor. However, if too many training patterns are provided, it will learn details and respond poorly to unseen patterns. In addition, neural networks often require a large set of training data. This proves to be impractical for real-life applications.

Genetic algorithms employ an artificial version of natural selection and use artificial genetic structures to solve problems. It is based on the theory of biological evolution in which characteristics of parents are transmitted to their offspring by means of genes that lead to the evolution of organisms. Grierson[103] proposed a coupling of neural network with genetic algorithms to arrive at an alternate best concept solution through evolution and artificial learning in the domain of bridge structure examples. Rafiq and Williams[102] also demonstrated how genetic algorithms could be coupled with artificial neural networks in the preliminary design of buildings. Taura *et al.*[104] proposed the use of genetic algorithm as part of the shape feature generating process model to aid in representing free-form shape features. The advantage of genetic algorithms is that they are capable of traversing large complex multi-dimensional search space to obtain a design solution. The major disadvantages are the difficulty in the selection of parameters and that they are slow.

Closely linked to neural networks are the machine learning techniques. Recent work in application of machine learning to engineering design can be found in Ref. 105–107. Archiszewski[105] is concerned with the adequacy of domain representation (irrelevant attributes, insufficient descriptors) which will allow reasoning by two methods — data-driven constructive induction, and hypothesis-driven constructive induction. O'Rorke[106] focuses on the abductive form of explanation-based learning. Case studies involving explaining physical processes, explaining decision, explaining signals are given. Rao[107] concentrates on using machine learning techniques to learn bi-directional models that can provide design synthesis support and hence reduce design iterations. This approach is aimed at parameterized domains (domains with fixed structure) so that all the variables in the design stage are well defined.

4. Current Scene

With the rapid increased in popularity of the World Wide Web, new research places more and more emphasis on the need to support collaborative design. Pahng et al.[108] proposed a framework for modeling and evaluating product design problems in a computer network-oriented design environment. In product design, many inter-related design decisions are made to meet potentially competing objectives. These decisions may span many disciplines. Thus, there is a need for an integrative framework that enables designers to rapidly construct performance models of complex problems and for information sharing electronically (e.g. via the internet). Approaching the same problem from a different angle, Ndumu and Tah[109] examined the use of agents to assist the design effort. An agent is a self-contained program capable of controlling its own decision-making and acting based on its perception of its environment, in pursuit of one or more objectives. Using two examples from construction supply chain provisioning and building design, the authors demonstrate the advantages that an agent-based approach brings to collaborative design. Tichkiewitch and Veron[110] propose yet another approach taken to aid in the integrated design environment. They present two models (the product model and the data model) and two exchange modes (a formal mode and an informal mode) to facilitate cooperative work between partners of the life-cycle of the product. Lu et al.[111] focus on the integration of various design and analysis models into a cohesive set to aid the collaborative negotiation process demanded by the design activity.

A separate yet distinct shift in the focus of design research is toward the support for virtual prototyping. Virtual prototyping refers to the analysis of a product without actually making a physical prototype of the product. Such analysis may be performed via the aid of expert system agents that reside in a distributed fashion on the internet. These agents require CAD information at different level of abstractions. Gadh and Sonthi[112] look into the different levels of geometric abstraction for achieving such virtual prototyping. Koch and Raczynski[113] looks into using virtual reality techniques as a supplement of the conventional CAD and rapid prototyping methods to achieve virtual prototyping. A digital mock-up is designed in virtual reality and is referred to as the virtual prototype. Drews and Weyric[114] focus on discussing the interaction and functional simulation of virtual prototyping. Dani and Gadh[115] use the virtual reality environment to allow the creation of concept shape designs rapidly. Cartwright[116] presents a modeling approach that allows the user to experiment, explore or make changes to the virtual prototypes.

5. Conclusion

While much progress has been made in the modeling and reasoning techniques to support conceptual design activity, there remains a large gap in transferring these techniques to real-life design applications. This is because many of the techniques face the problem of scaling-up. In addition, some of these techniques make certain simplifying assumptions that are unrealistic for real-world applications. For example, in the case of using neural networks to map functions to forms, the implicit assumption is that there exists a one-to-one mapping from functions to forms. This is certainly not the case since many different forms can realize one function and one form can satisfy multiple functions. Thus, a large part of conceptual design activity still depends largely on the creative abilities of the human designer.

In this paper, we have performed a preliminary survey of the tools and techniques that have been proposed to aid in the conceptual design of mechanical products. The survey results show that in spite of the great advances in both the modeling and reasoning techniques much remains to be done. We hope this survey will serve to motivate researchers to look closely at the underlying modeling and reasoning techniques for conceptual design of mechanical products, and perhaps to derive an integrated framework for the next generation of computer-aided design tools.

References

1. D. Neville and L. Joskowicz, A representation language for mechanical behavior, Design theory and methodology, *ASME* **53** (1993) 1–6.
2. S. Mullins and J. R. Rinderle, Grammatical approaches to engineering design, Part I: An introduction and commentary, *Research in Engineering Design* **2** (1991) 121–135.
3. J. Rinderle, Grammatical approaches to engineering design, Part II: Melding configuration and parametric design using attribute grammars, *Research in Engineering Design* **2** (1991) 137–146.
4. M. Vescovi, Y. Iwasaki, R. Fikes and B. Chandrasekaran, CFRL: A language for specifying the causal functionality of engineered devices, *AAAI-93* (1993).
5. C. Carlson, Grammatical programming: An algebraic approach to the description of design spaces, Pittsburgh, Carnegie Mellon University, 1993.
6. G. Stiny, Introduction to shape and shape grammars, *Environment and Planning B: Planning and Design* **7** (1980) 343–351.
7. J. Heisserman, Generative geometric design and boundary solid grammars, Pittsburgh, Carnegie Mellon University, 1991.

8. W. J. Mitchell, Articulate design of free-form structures, Artificial intelligence in structural engineering: Information technology for design, collaboration, maintenance and monitoring, 1998, 223–234.

9. G. Reddy and J. Cagan, Optimally directed Truss topology generation using shape annealing, *ASME Design Automation Conference*, Albuquerque, NM, 1993.

10. L.C. Schmidt and J. Cagan, Recursive annealing: A computational model for machine design, Design theory and methodology, *ASME*, 1993, 243–251.

11. S. Szykman and J. Cagan, Automated generation of optimally directed three dimensional component layouts, *ASME Design Automation Conference*, Albuquerque, NM, 1993.

12. K. N. Brown, J. H. Sims Williams and C. A. McMahon, *Grammars of Features in Design* (Kluwer, 1992).

13. S. N. Longnecker and P. A. Fitzhorn, A shape grammar for non-manifold modeling, *Research in Engineering Design* **2** (1991) 159–170.

14. E. Tyugu, Attribute models of design objects, IFIP transactions: Formal design methods for CAD, 1994, 33–34.

15. K. Andersson, P. Makkonen and J. G. Persson, A proposal to a product modelling language to support conceptual design, *Annals of CIRP* **44**, 1 (1995) 129–132.

16. L. K. Alberts, YMIR: A sharable ontology for the formal representation of engineering design knowledge, IFIP transactions: Formal design methods for CAD, 1994, 3–32.

17. M. R. Cutkosky, R. S. Engelmore, R. E. Fikes, M. R. Genesereth, T. R. Gruber, W. Mark and J. M. Tenenbaum, PACT: An experiment in integrating concurrent engineering systems, *IEEE Computer*, 1993.

18. B. Kuipers, Commonsense reasoning about causality: Deriving behavior from structure, *Artificial Intelligence* **24** (1984) 169–203.

19. J. Malmqvist, Computer-aided conceptual design of energy transforming technical systems based on technical system theory and bond graphs, *Lancaster International Workshop on Engineering Design*, 1994, 59–78.

20. S. Murthy and S. Addanki, PROMPT: An innovative design tool, *Proceedings AAAI-87*, 1987, 637–642.

21. S. Joshi and T. C. Chang, Graph-based heuristic for recognition of machined features from a 3D solid model, *Computer-Aided Design* **20** (1988) 58–66.

22. J. C. H. Chung and M. D. Schussel, Comparison of variational and parametric design, *Autofact-89*, 1989.

23. A. Kusiak and N. Larson, Decomposition and representation methods in mechanical design, *Journal of Mechanical Design* **117** (1995) 17–24.

24. A. Kusiak and E. Szczerbicki, A formal approach to specifications in conceptual design, *Journal of Mechanical Design* **114** (1992) 659–666.

25. R. Arnheim, *Visual Thinking, USA* (University of California Press, 1969).

26. R. H. McKim, *Experiences in Visual Thinking* (MA, PWS Engineering, USA, 1980).

27. D. G. Ullman, S. Wood and D. Craig, The importance of drawing in the mechanical design process in preprints of NSF engineering design research conference, University of Massachusetts, USA, 1989.

28. K. Ehrlenspiel and N. Dylla, Experimental investigation of the design process, *ICED-89*, 1989, 77–98.

29. D. F. Radcliffe and T. Y. Lee, Models of visual thinking by novice designers, Design theory and methodology, *ASME*, 1990, 145–152.

30. E. Sittas, 3D design reference framework, *Computer-Aided Design* **23**, 5 (1991) 380–384.

31. J. Koski, Multicriteria optimization in structural design: State of the art, *Advances in Design Automation, ASME*, 1993, 621–929.

32. R. R. Levary, Engineering applications of operations research, *European Journal of Operational Research* **72** (1994) 32–42.

33. F. L. Krause and J. Schlingheider, Development and design with knowledge-based software tools — An overview, expert systems with applications **8**, 2 (1995) 233–248.

34. A. A. G. Requicha, Representations for rigid solids: Theory, methods and systems, *Computing Surveys* **12**, 4 (1980) 437–464.

35. J. J. Weiler, Topological structure for geometric modeling, Troy, New York, Rensselaer Polytechnic Institute, 1986.

36. V. C. Lin, Variational geometry in computer-aided design, Massachusetts Institute of Technology, MA, 1981.

37. R. A. Light and D. C. Gossard, Modification of geometric models through variational geometry, *Computer-Aided Design* **14**, 4, 1982.

38. M. Shpitalni and H. Lipson, Automated reasoning for design under geometrical constraints, *Annals of CIRP* **46**, 1 (1997) 85–88.

39. J. R. Dixon, Designing with features for component design, *Workshop on Features in Design and Manufacturing*, National Science Foundation, 1988.

40. E. C. Libardi and J. R. Dixon, Designing with features: Design and analysis of castings as an example, *Proceedings Spring National Design Conference*, 1986.

41. N. Nakajima and D. Gossard, Basic study on feature descriptor, Cambridge, Massachusetts Institute of Technology, MA, 1982.

42. S. C. Luby, J. R. Dixon and M. K. Simmons, Creating and using a feature data base, *Computer in Mechanical Engineering*, 1986.

43. M. J. Pratt, Solid modeling and the interface between design and manufacturing, *IEEE Computer Graphics and Application*, 1984, 52–59.

44. D. Roller, Design by features: An approach to high level shape manipulations, *Computers in Engineering* **12** (1989) 185–191.

45. J. Shah, A. Bhatnagar and D. Hsiao, Feature mapping and application shell, *Computers in Engineering* **1** (1988) 489–496.

46. M. Dyer, M. Flowers and J. Hodges, An engineering design invention system operating naively, *Proceedings of the 1st Conference of Applications of Artificial Intelligence in Engineering Problems*, 1986, 327–341.

47. M. R. Cutkosky, J. M. Tenenbaum and D. Muller, Features in process-based design, *Proceedings International Computers in Engineering Conference, ASME*, 1988.

48. C. C. Hayes and P. K. Wright, Automating process planning: Using feature interactions to guide search, *Journal of Manufacturing Systems* **8**, 1, 1989.

49. R. Gadh, A hybrid approach to intelligent geometric design using features-based design and feature recognition, *Advances in Design Automation, ASME*, 1993, 273–283.

50. J. Han and A. A. G. Requicha, Integration of feature based design and feature recognition, *Computer-Aided Design* **29**, 5 (1997) 393–403.

51. W. Keirouz, J. Pabon and R. Young, Integrating parametric geometry, features and variational modeling for conceptual design, *Design Theory and Methodology*, 1990, 1–9.

52. P. A. van Elsas and J. S. M. Vergeest, Displacement feature modelling for conceptual design, *Computer-Aided Design* **30**, 1 (1998) 19–27.

53. A. V. Rao and N. S. Prakasa, BEAS: Expert system for the preliminary design of bearings, *Advances in Engineering Software* **14** (1992) 163–166.

54. C. Tong and A. Gomory, A knowledge-based computer environment for the conceptual design of small electromechanical appliances, *Computer* **26**, 1 (1993) 69–71.

55. P. Arpaia, G. Betta, A. Langella and M. Vanacore, Expert system for the optimum design of measurement systems, *IEEE Proceedings on Scientific Measurement Technology* **142**, 4 (1995) 330–336.

56. D. Carstoiu, G. Dobre, C. Grigorescu and S. Grigorescu, A new approach to expert systems design, *Advances in Intelligent Systems*, 1997, 520–525.

57. A. Taleb-Bendiab, ConceptDesigner: A knowledge-based system for conceptual engineering design, *Proceedings 9th International Conference on Engineering Design*, 1993, 1303–1311.

58. R. H. Bracewell, R. V. Chaplin, P. M. Langdon, M. Li, V. K. Oh, J. E. E. Sharpe and X. T. Yan, Integrated platform for AI support of complex design: Rapid development of schemes from first principles, AI system support for conceptual design, *Proceedings of the 1995 Lancaster International Workshop in Engineering Design* (Springer-Verlag, 1995) 170–188.

59. T. Kiriyama, K. Kurumatani, T. Tomiyama and H. Yoshikawa, Metamodel: An integrated modeling framework for intelligent CAD, *Artificial Intelligence in Design*, 1989, 429–449.

60. Q. J. Harmer, P. M. Weaver and K. M. Wallace, Design-led component selection, *Computer-Aided Design* **30**, 5 (1998) 391–405.

61. M. A. Kolb and M. W. Bailey, FRODO: Constraint-based object-modeling for preliminary design, *Advances in Design Automation, ASME*, 1993, 307–318.

62. J. S. Oh, P. O'Grady and R. E. Young, A constraint network approach to design for assembly, *IIE Transactions* **27** (1995) 72–80.

63. C. J. Moore, J. C. Miles and J. V. Cadogan, Applying quantitative constraint satisfaction in preliminary design, Artificial intelligence in structural engineering: Information technology for design, collaboration, maintenance and monitoring, 1998, 235–248.

64. R. Vujosevic, A. Kusiak and E. Szczerbicki, Reason maintenance in product modeling, *Journal of Engineering for Industry* **117** (1995) 223–231.

65. Z. Yao and A. L. Johnson, On estimating the feasible solution space of design, *Computer-Aided Design* **29**, 9 (1997) 649–655.

66. A. K. Goel and B. Chandrasekaran, Case-based design: A task analysis, *Artificial Intelligence in Engineering Design*, eds. C. Tong and D. Sriram (Academic Press, 1992).

67. K. P. Sycara and D. Navinchandra, Retrieval strategies in a case-based design system, *Artificial Intelligence in Engineering Design* (Academic Press Inc., 1992) 145–163.

68. C. L. Li, S. T. Tan and K. W. Chan, A qualitative and heuristic approach to the conceptual design of mechanisms, *Engineering Applications in Artificial Intelligence* **9**, 1 (1996) 17–31.

69. P. Gomes and C. Bento, A case-based approach for elaboration of design requirements, Case-based reasoning: Research and development, *Proceedings of the Second International Conference, ICCBR-97*, 1997, 33–42.

70. W. Hsu, A. Lim and C. S. G. Lee, Conceptual level design for assembly analysis using state transitional approach, *Proceedings of the IEEE International Conference of Robotics and Automation*, 1996, 3355–3361.

71. C. S. Irgens, Design support based on projection of information across the product-development life cycle by means of case-based reasoning, *IEE Proceedings — Science Measurement Technology* **142**, 5 (1995) 345–349.

72. M. Simina and J. Kolodner, Creative design: Reasoning and understanding, case-based reasoning: Research and development, *Proceedings of the Second International Conference, ICCBR-97*, 1997, 587–598.

73. J. Mostow, M. Barley and T. Weinrich, Automated reuse of design plans in BOGART, *Artificial Intelligence in Engineering Design* (Academic Press, 1992).

74. R. Banares-Alcantara, J. M. P. King and G. H. Ballinger, EGIDE: A design support system for conceptual chemical process design, AI system support for conceptual design, *Proceedings of the 1995 Lancaster International Workshop on Engineering Design* (Springer-Verlag, 1995) 138–152.

75. T. Murakami and N. Nakajima, Mechanism concept retrieval using configuration space, *Research in Engineering Design* **9** (1997) 99–111.

76. J. L. Martin and W. M. K. Roddis, Integrating qualitative and quantitative reasoning in structural engineering, *Computing in Civil and Building Engineering* **2** (1993) 1235–1242.

77. M. Ohki, H. Shinjo and M. Abe, Design support to determine the range of design parameters by qualitative reasoning, *IEEE Transactions on Systems, Man and Cybermetics* **24**, 5 (1994) 813–819.

78. M. Yoshioka, M. Nakamura, T. Tomiyama and H. Yoshikawa, A design process model with multiple design object models, Design theory and methodology, *ASME* **53** (1993) 7–14.

79. R. Nacaneethakrishnan, K. L. Wood and R. H. Crawford, An object-oriented formalism for geometric reasoning in engineering design and manufacture, *Advances in Design Automation, ASME*, 1993, 301–313.

80. A. Kusiak, E. Szczerbicki and R. Vujosevic, An intelligent system for conceptual design, *Expert Systems* **3**, 2 (1991) 35–44.

81. J. Bento, B. Feijo and D. L. Smith, Engineering design knowledge representation based on logic and objects, *Computers and Structures* **63**, 5 (1997) 1015–1032.

82. J. D. Kleer and J. S. Brown, A qualitative physics based on confluences, *Artificial Intelligence* **24** (1984) 7–83.

83. B. Faltings, E. Baechlet and J. Primus, Reasoning about kinematic topology, *Proceedings IJCAI-89*, 1989, 1331–1336.

84. K. Kurumatani, T. Tomiyama and H. Yoshikawa, Qualitative representation of machine behaviours for intelligent CAD systems, *Mechanical Machine Theory* **25**, 3 (1990) 325–334.

85. K. D. Forbus, Qualitative process theory, *Artificial Intelligence* **24** (1984) 84–168.

86. J. Hodges, Naive mechanics: A computational model of device use and function in design improvisation, *IEEE Expert* **7**, 1 (1992) 14–27.

87. K. W. Roddis and J. L. Martin, Qualitative reasoning about steel bridge fatigue and fracture, *Fifth International Workshop on Qualitative Reasoning about Physical Systems*, 1991.

88. D. I. Schwartz and S. S. Chen, Spatial and temporal aspects of qualitative structural reasoning, *Proceedings of the 8th ASCE Conference on Computing in Civil Engineering*, 1992.

89. L. M. Bozzo and G. L. Fenves, Qualitative evaluation of preliminary structural design, *Proceedings of the 8th ASCE Conference on Computing in Civil Engineering*, 1992.

90. R. Fruchter, Deriving alternative structural modifications based on qualitative interpretation at the conceptual design stage, *Computing in Civil and Building Engineering* **2** (1993) 1243–1250.

91. R. Simmons, Commonsense arithmetic reasoning, *Proceedings of AAAI 86*, 1986.

92. J. Kalagnanam, M. Henrion and E. Subrahmanian, The scope of dimensional analysis in qualitative reasoning, *Computational Intelligence* **10**, 2 (1994) 117–133.

93. D. I. Schwartz and S. S. Chen, A constraint-based approach for qualitative matric structural analysis, *Artificial Intelligence for Engineering Design, Analysis and Manufacturing* **9** (1995) 23–36.

94. S. Parsons, Interval algebra and order of magnitude reasoning, *Applications of AI in Engineering* **vi** (1991) 945–961.

95. Q. Shen and R. Leitch, Combining qualitative simulation and fuzzy sets, *Recent Advances in Qualitative Physics*, eds. B. Faltings and P. Struss (MIT Press, 1992).

96. A. Gelsey, Automated reasoning about machine geometry and kinematics, *Proceedings of the 3rd IEEE Conference on AI Applications*, 1987.

97. K. D. Forbus, P. Nielsen and B. Faltings, Qualitative spatial reasoning: The clock project, *Artificial Intelligence* **51** (1991) 417–471.

98. K. H. Law, Some personal experience in computer aided engineering research, Artificial intelligence in structural engineering: Information technology for design, collaboration, maintenance and monitoring, 1997, 178–195.

99. R. P. Lippmann, An introduction to computing with neural nets, *IEEE ASSP Magazine*, 1987, 4–22.

100. S. R. T. Kumara and S. V. Kamarthi, Application of adaptive resonance networks for conceptual design, *Annals of the CIRP* **41**, 1 (1992) 213–216.

101. S. L. Hung and H. Adeli, Object-oriented backpropagation and its' application to structural design, *Neurocomputing* **6** (1994) 45–55.

102. M. Y. Rafiq and C. Williams, An investigation into the integration of neural networks with the structured genetic algorithm to aid conceptual design, Artificial intelligence in structural engineering: Information technology for design, collaboration, maintenance and monitoring, 1998, 295–307.

103. D. E. Grierson, Conceptual design using evolutive-cognitive techniques, *Computing in Civil Engineering* **2** (1994) 2183–2190.

104. T. Taura, I. Nagasaka and A. Yamagishi, Application of evolutionary programming to shape design, *Computer-Aided Design* **30**, 1 (1998) 29–35.

105. T. Archiszewski, E. Bloedorn, R. S. Michalski, M. Mustafa and J. Wnek, Machine learning of design rules: Methodology and case study, *Journal of Computing in Civil Engineering* **8**, 3 (1994) 286–308.

106. P. O'Rorke, Abduction and explanation-based learning: Case studies in diverse domains, *Computational Intelligence* **10**, 3 (1994) 295–330.

107. R. B. Rao and S. C. Y. Lu, Inverse engineering: A methodology for learning models to support engineering design, *IEEE 9th Conference on Artificial Intelligence for Applications*, 1993.

108. F. Pahng, N. Senin and D. Wallace, Distribution modeling and evaluation of product design problems, *Computer-Aided Design* **30**, 6 (1998) 411–423.

109. D. T. Ndumu and J. M. H. Tah, Agents in computer-assisted collaborative design, Artificial intelligence in structural engineering: Information technology for design, collaboration, maintenance and monitoring, 1998, 249–270.

110. S. Tichkiewitch and M. Veron, Methodology and product model for integrated design using a multiview system, *Annals of the CIRP* **46**, 1 (1997) 81–84.

111. S. C.-Y. Lu, D. Li, J. Cheng and C. L. Wu, A model fusion approach to support negotiations during complex engineering system design, *Annals of the CIRP* **46**, 1 (1997) 89–92.

112. R. Gadh and R. Sonthi, Geometric shape abstractions for internet-based virtual prototyping, *Computer-Aided Design* **30**, 6 (1998) 473–486.
113. R. Koch and A. Raczynski, Virtual prototyping for geometry based product development, *Proceedings of IEEE 24th Annual Conference of IEEE Industrial Electronics, IECON-98* **4** (1998) 2158–2161.
114. P. Drews and M. Weyrich, Interactive functional evaluation in virtual prototyping illustrated by an example of a construction machine design, *Proceedings of IEEE 24th Annual Conference of IEEE Industrial Electronics, IECON-98* **4** (1998) 2143–2145.
115. T. H. Dani and R. Gadh, Creation of concept shape designs via a virtual reality interface, *Computer-Aided Design* **29**, 8 (1997) 555–563.
116. A. J. Cartwright, Interactive prototyping — A challenge for computer based design, *Research in Engineering Design* **7** (1997) 10–19.

CHAPTER 2

COMPUTER TECHNIQUES AND APPLICATIONS IN RAPID PROTOTYPING IN MANUFACTURING SYSTEMS

T. W. LAM

Department of Mechanical Engineering,
The Hong Kong Polytechnic University,
Hong Kong

K. M. YU

Department of Mechanical Engineering,
The Hong Kong Polytechnic University,
Hong Kong

C. L. LI

Department of Manufacturing Engineering and Engineering Management,
City University of Hong Kong,
Hong Kong

Rapid prototyping processes, like stereolithography, have gained wide industrial acceptance in recent years. The main advantage of rapid prototyping over conventional manufacturing processes is faster speed. Unlike material removal from a stock (as in machining), most rapid prototyping processes have to build the solid volume layer by layer. The tracing of the cross-sectional solid area in each layer is the most time consuming process. The process can be speeded up if the material volume to be traced can be reduced by extracting empty volumes in the original solid. With a Constructive Solid Geometry approach, the hollow solid generated by employing the negative solid offset technique is subjected to the problem of no interior support. A sub-boundary octree approach of generating thin shell solids with reinforced interior structuring is then proposed to re-solve the problem. As a result, the combination of the negative solid offset technique and the sub-boundary octree approach is proved to be success in generating thin shell solid with reinforced structuring that can be traced in a much faster speed. Algorithms have been implemented to demonstrate the validity of the proposed theory. Various testing models are produced by a Fused Deposition Modelling rapid prototyper and the results are supportive. The proposed theory can generate prototypes with good dimensional accuracy and surface flatness.

Keywords: Hollow rapid prototype; thin-shell rapid prototype; offset; octree; fused deposition modelling.

1. Introduction

In recent years, many rapid prototyping (RP) machines were sold and in use world-wide. RP is, however, a new technology. Many problems still exist and are waiting to be solved.[2,3] Simply speaking, RP can be considered as a three-dimensional printing process. It converts the geometric data in a computer-aided design (CAD) model to a three-dimensional physical prototype. The engineers can check the form, fit and function of the prototype rather than examining it on the monitor screen. Whether a prototype is suitable for function checking depends, however, on the material being used. The process is "rapid" because a prototype can be completed in hours instead of days, weeks, or months when using the traditional methods. The comparison includes the tooling and part programming time in the proto-type development cycle.[3,4] Traditional prototyping performs the job by machining and/or hand-carving clay, wood or foam. There are various RP techniques available in the market. Some popular machines used are based on the principles of one of the following: stereo-lithography (SLA),[2,5] photo-masking, selective laser sintering (SLS),[6] fused deposition modelling (FDM),[7] laminated object manufacturing (LOM),[8] "drop-on-demand" inkjet process (DODJet),[9] and three-dimensional printing (3DP).[10]

For instance, the SLA machine shown in Fig. 1 creates the prototype by tracing layer cross-section on the surface of the liquid photo-polymer pool with a Helium-Cadmium ultra-violet laser beam. Unlike the contouring or zig-zag cutter movement being used in CNC machining, the beam traces in parallel lines first in one direction and then in the orthogonal direction using a patented technique called STAR-WEAVE. The beam will penetrate the previous layer to fuse the new one on

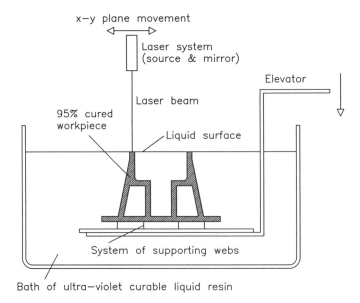

Fig. 1.   Working principle of SLA.

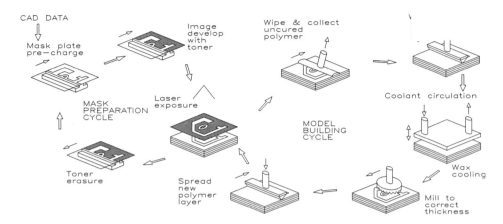

Fig. 2. Working principle of photo masking.

top. The platform will then submerge the prototype down a user-specified increment and a wiper will help to spread the viscous polymer over for building the next layer. When all layers is completed, the prototype is still about 95% cured. Post-curing is needed to convert all resin to solid plastic. This is done in another fluorescent equipment, called post-curing apparatus (PCA).

The photo-masking technique shown in Fig. 2 involves four main steps: photo-masking, wax-filling, curing and milling. For each layer, the machine creates a new mask on a glass plate, the process similar to photocopying. The plate is shuttled over a resin-coated workpiece for the ultra-violet laser exposure. The excess liquid resin uncured will then be wiped off and a liquid wax layer is coated over the work. After the wax is cooled to solid, a milling cutter will cut to expose the "true" part surface thickness between each layer. When all layers are finished, the wax block that support the part during the process will be melted away. As in SLA, the process uses liquid polymer and more than one part can be made in one go. Unlike the vector scanning being used in SLA, a solid area is cured in each laser exposure. It is thus faster and does not need any post-curing.

The SLS machine shown in Fig. 3 traces the cross-section on a powder surface with carbon dioxide laser to create a sintered layer. Note that sinters means softens and bonds, and it will not melt the powder. The whole platform is lowered down and a roller will then level fresh powder over the part surface for the next pass. The unfused powder will not be removed and will serve as the support of the part during the building process.

The FDM process shown in Fig. 4 builds a part by extruding melted resin, proprietary nylon or wax from a computer-controlled heated head. The material is fed as a continuous filament. The head traces an exact outline of each cross-section layer of the part design. The nozzle is then raised for the next layer. This method has minimum material wastage.

The DODJet process shown in Fig. 5 fabricates the prototype by depositing both thermoplastic and wax materials on the build substrate. The build and support

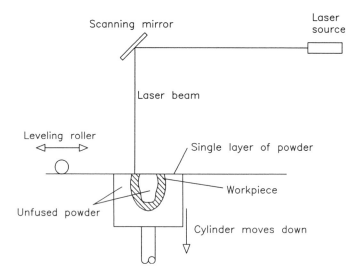

Fig. 3.   Working principle of SLS.

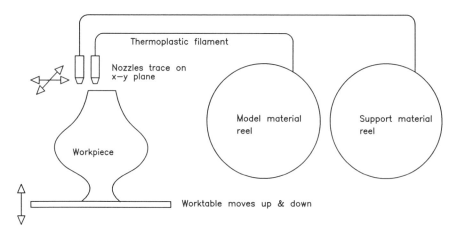

Fig. 4.   Working principle of FDM.

materials are deposited in a series of micro-droplets. After the liquid-to-solid phase transition, the droplets adhere together to form the prototype.

The 3DP process shown in Fig. 6 consists of five steps: collect powder, spread powder, discharge excess powder, print by depositing binder, and lowering the build box by one layer thickness. The binder is printed onto each layer of powder to form the shape of the cross-section of the prototype. After lowering the build box by one layer thickness, a new layer of powder is spread. The five-step process is repeated until the build is complete. Once a build is complete, the excess powder is vacuumed away to expose the prototype.

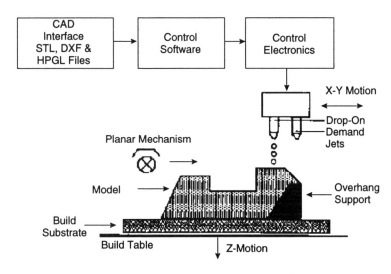

Fig. 5. Working principle of DODJet.[9]

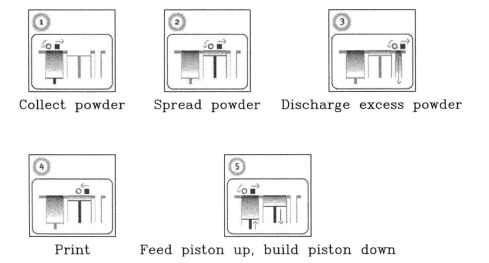

Fig. 6. Working principle of 3DP.[10]

In summary, the methods work under the same principle of "growing" instead of material removal. Layers of material cross-section will grow bottom-up to produce the final shape. The information flow of RP is:

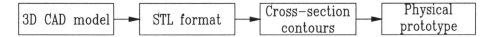

A three-dimensional CAD model of either wireframe, surface or solid is first converted into an STL format.[11] Conversion from a solid model will be the most reliable as it contains the most complete geometrical information. STL format is the now commonly-used "standard" format in inputting to RP machines (STL format only has the status of AutoCAD DXF but not IGES,[12] the latter is a true standard.) It was originally used in stereo-lithography apparatus (SLA).[11] The format contains a list of triangles (vertex coordinates and outward normal) that describes the boundary of the object to be built (similar to those used in ray-casting objects faceted by triangles). A new standard is under discussion between the various vendors as STL is bulky in size (e.g. outward normal is a redundant information) and inefficient to process.

From Figs. 1–4 and Refs. 13 & 14, one can see that most RP techniques require the tracing of solid cross-sectional area. LOM is the exception, while photo-masking still needs cutter movement to machine the cross-section. The tracing by a laser beam (SLA, SLS) or polymer filament (FDM) is in fact the time-critical step of the RP process. The process can thus be accelerated if some way can be devised to speed up the tracing, or the stuff to be traced can be reduced. Speeding up the tracing is a hardware problem with the RP machine. In this chapter, the problem is attacked by using the latter method, i.e. reducing the material volume to be traced. One may consider that the reduction process is achieved by leaving voids or empty spaces in the original solid interior. This is justified on the assumption that the prototype is used for form and fit examination. If the prototype with reduced material volume is strong enough, it may also be used for function checking. Therefore, if the boundary shape and size of the reduced-volume prototype remains unchanged, form and fit checking should give identical results to the original one. Since the reduced-volume solid has less cross-sectional solid area to fill, this also implies that less laser power will be consumed and less material will be wasted (depending on the type of RP, some material trapped can be recycled while some may not). Also, a prototype with less volume will be lighter in weight and a cheaper RP machine can be used to build it.

There are two possible ways to reduce the volume to be traced. One is to generate a reduced volume model from the 3D CAD model to be input to the RP machine. Another is to produce reduced cross-section contours from the solid cross-sections prior to tracing. The two options are as shown below:

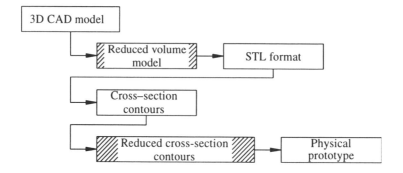

However, the RP machine vendors have not disclosed their interfaces from cross-section slicing to prototype tracing. It is therefore more convenient and flexible for the design engineer to pre-process the CAD model rather than to post-process the cross-sections.

The reduction can be achieved by leaving some empty volumes in the original solid. The empty volumes can be generated by negative offsetting[1] the original solid so that the new solid has the volume of the original solid minus the offset solid. One advantage is that offset can be done automatically and the user needs only to specify the offset distance. In addition, offset is relatively well studied since the use of numerical control machining. For instance, normal offset has been used to find cutter location data in Refs. 15 & 16. Mathematical theories and implementation algorithms for offset are numerous.[1,17–24] The offset result will be a thin shell solid with new internal structure. The proofs of using negative solid offset to create thin shell solid, its implementation and practical limitations are discussed in Sec. 2. To rectify the practical limitations, a sub-boundary octree approach for producing a thin shell rapid prototype with reinforced interior structure is proposed. The octree reinforced thin shell theory and its implementation are presented in Sec. 3. Finally, case studies of thin shell RP are illustrated in Sec. 4.

2. Thin Shell Solid

The thin shell solid, i.e. the original solid differences the offset solid, should have the properties that (a) the volume of material (the interior) is reduced and (b) the original solid boundary (the exterior) is unchanged. The theory is explained in terms of Constructive Solid Geometry (CSG)-represented solids.

2.1. *Solid offset*

Solid offset is rigorously defined in Ref. 1. Positive offset will result in larger volume. Thus, the offset direction of the (interior and exterior shell) boundary is away from the material side. Negative offset will result in smaller volume. Thus, the offset direction of the boundary is towards the material side. For implementation details of offset algorithms and techniques to handle tricky cases (like self-intersection, rounded corners, non-uniform wall thickness, etc.), Refs. 1, 17, 18, 23 & 24 may be consulted, though not all of them are for solid offset. In what follows, mathematical proofs for the set operators of union, difference and intersection will be given first. The conversion for a CSG-represented solid will then be explained.

2.2. *Offset of union*

According to Ref. 1, the conversion of a simple unioned CSG solid to its reduced-volume counterpart by solid offset is (Fig. 7(a))

$$A \cup^* B \to (A \cup^* B) -^* (A^- \cup^* B^-).$$

The aim is to prove that both properties (a, b) are satisfied in the conversion result.

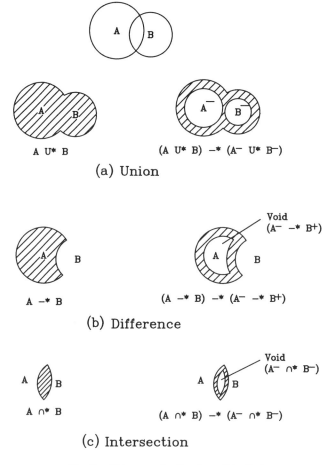

Fig. 7.   Volume reduction by solid offset.

Proof:

(a) To prove that the conversion result has a reduced volume, one only needs to prove that the conversion result is a sub-set of the original solid.

Obviously, $(A \cup^* B) -^* (A^- \cup^* B^-) \subseteq A \cup^* B$ is true in general. Thus, the volume is reduced. Note also that the improper set inclusion is used since A^- will be a null set when the offset distance is too large (Fig. 8). (A^+ will only be a null set if A is null.) Hence, one may include checking of non-null A^- and B^- in actual implementation.

(b) In the conversion, the boundary of the original solid should remain intact. However, new boundary will also be generated by solid offset in the conversion. If the two boundaries do not intersect then the conversion will satisfy the purposes of both volume reduction and integrity of the original boundary. Mathematically, this is to check for non-intersection of boundaries of $(A \cup^* B)$ and $(A^- \cup^* B^-)$. In general, one needs only to prove that the boundary of $(A^- \cup^* B^-)$ is inside

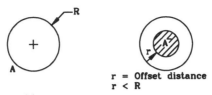

(a) Non−null negative offset result.

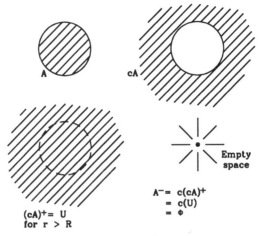

(b) Null negative offset result (offset distance too large).

Fig. 8. Non-null and null negative offset results.

(i.e. a sub-set of) $(A \cup^* B)$. The proof is shown algebraically as follows:

$$\partial(A^- \cup^* B^-) \subset A^- \cup^* B^-$$
$$\subseteq (A \cup^* B)^{-1}$$
$$\subset i(A \cup^* B).$$

Hence, $\partial(A \cup^* B) \cap \partial(A^- \cup^* B^-) = \emptyset$, or the offset boundary will not intersect with the original boundary.

2.3. *Offset of difference*

The conversion of a simple differenced CSG solid to its reduced-volume counterpart by solid offset[1] is (Fig. 7(b))

$$A -^* B \rightarrow (A -^* B) -^* (A^- -^* B^+).$$

Proof:

(a) $(A -^* B) -^* (A^- -^* B^+) \subseteq A -^* B$ is true in general. Thus, the volume is reduced.

(b) $\partial(A^- -^* B^+) \subset A^- -^* B^+$
$$= (A -^* B)^{-1}$$
$$\subset i(A -^* B).$$
Hence, $\partial(A -^* B) \cap \partial(A^- -^* B^+) = \emptyset$.

2.4. Offset of intersection

The conversion of a simple intersect CSG solid to its reduced-volume counterpart by solid offset[1] is (Fig. 7(c))

$$A \cap^* B \to (A \cap^* B) -^* (A^- \cap^* B^-).$$

Proof:

(a) $(A \cap^* B) -^* (A^- \cap^* B^-) \subseteq A \cap^* B$ is true in general. Thus, the volume is reduced.

(b) $\partial(A^- \cap^* B^-) \subset A^- \cap^* B^-$
$$= (A \cap^* B)^- {}^1$$
$$\subset i(A \cap^* B).$$

Hence, $\partial(A \cap^* B) \cap \partial(A^- \cap^* B^-) = \emptyset$.

2.5. Offset of CSG solid

CSG solid is defined by the combination of solid primitives through binary regularized set operators (i.e. regularized set operators with two operands). Common solid primitives include block, sphere, cylinder, cone (or frustum), wedge (and tetrahedron) and torus. The positive and negative solid offset of the solid primitives can be easily computed by changing the sizes and/or origin position of the original primitives. (In our application, a constant-radius offset for cone, that will round the apex, is not necessary. A simple shift of reference origin will do the job.) Hence, the reduced-volume solid can be obtained by applying the conversion rules just mentioned to the original CSG-represented solid. The mathematical justification is as follows.

Let S, S^- be the CSG solid and its negative offset counterpart. The conversion is $S \to S -^* S^-$. Obviously, the conversion result, $S -^* S^-$, has a reduced volume (i.e. $S -^* S^- \subseteq S$). In addition, the boundary of the negative offset will not interfere with the boundary of the original solid as $\partial S^- \subset S^- \subset iS$ or $\partial S \cap \partial S^- = \emptyset$.

This is better illustrated with an example as follows (Fig. 9):

If $S = ((A \cup^* B) -^* C) \cap^* D$ where S is CSG defined solid, A, B, C and D are primitive volumes, then the procedure to reduce S, i.e. $S \to S -^* S^-$, is

$$
\begin{aligned}
S -^* S^- &= (((A \cup^* B) -^* C) \cap^* D) -^* (((A \cup^* B) -^* C) \cap^* D)^- \\
&= (((A \cup^* B) -^* C) \cap^* D) -^* (((A \cup^* B) -^* C) \cap^{*-} D) \\
&= (((A \cup^* B) -^* C) \cap^* D) -^* (((A \cup^* B) -^* C)^- \cap^* D^-) \\
&= (((A \cup^* B) -^* C) \cap^* D) -^* (((A \cup^* B) -^{*-} C) \cap^* D^-) \\
&= (((A \cup^* B) -^* C) \cap^* D) -^* (((A \cup^* B)^- -^* C^+) \cap^* D^-) \\
&= (((A \cup^* B) -^* C) \cap^* D) -^* (((A \cup^{*-} B) -^* C^+) \cap^* D^-) \\
&= (((A \cup^* B) -^* C) \cap^* D) -^* (((A^- \cup^* B^-) -^* C^+) \cap^* D^-).
\end{aligned}
$$

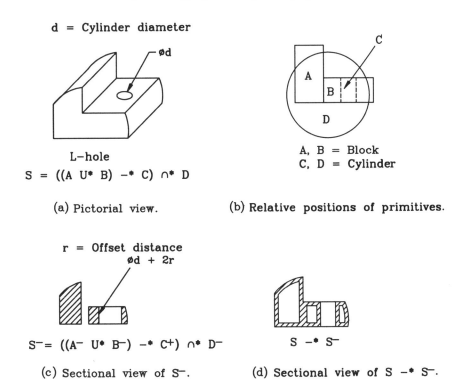

d = Cylinder diameter

L-hole

$S = ((A \; U^* \; B) \; -^* \; C) \cap^* D$

(a) Pictorial view.

A, B = Block
C, D = Cylinder

(b) **Relative positions of primitives.**

r = Offset distance
∅d + 2r

$S^- = ((A^- \; U^* \; B^-) \; -^* \; C^+) \cap^* D^-$

(c) Sectional view of S^-.

$S \; -^* \; S^-$

(d) **Sectional view of** $S \; -^* \; S^-$.

Fig. 9.   Negative offset for a simple part.

2.6. *Implementation*

This section will discuss the implementation issues of thin shell solid generation. The data structures being used are first described. The algorithms to generate offsets and subtracting them from the original solid are then explained.

2.6.1. *Data structures for void making*

As CSG tree is a binary tree with each node having two children, a series of linked data structures can be used to store the nodal information. The information includes the relationship between the parent node and the child node, the primitive type, the Boolean operation, etc. The information being used in the implementation are:

(1) Object Primitive Type
    It tells us what kind of primitive the child is, or whether it is a composite.
(2) Object Primitive Union
    Under this data structure, there is a Boolean operation data structure that tells us the type of Boolean operation being used to combine its two child nodes. The information about the child node can be addressed from the child node

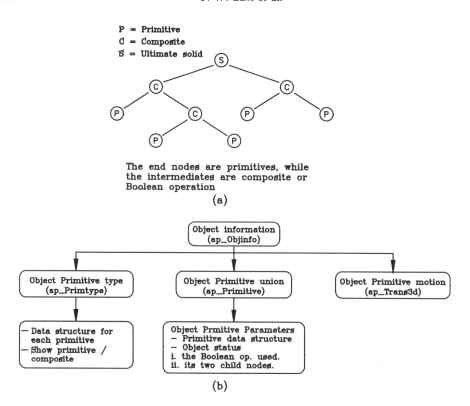

The end nodes are primitives, while
the intermediates are composite or
Boolean operation

(a)

(b)

Fig. 10.   (a) A CSG tree example; (b) Nodal information data structure.

data structure. This data structure also includes the parameter values for each kind of primitive.

(3) Object Information

It covers the above data structures plus the rigid motion matrix data of the object.

The inter-relationship of the above information is shown in Fig. 10.

2.6.2. *Algorithms for void making*

First of all, an algorithm should be designed to retrieve all child node information for the solid (object) concerned. The information will be used to generate a smaller object offset from the original one by increasing or decreasing the dimension values of the primitive nodes. The hollow solid is then obtained by subtracting the offset solid from the original object. Although all nodes in the CSG tree are stored in the same format, the span and size of the CSG tree are different for different objects. In fact, the same object may have more than one CSG representation.

Second, we need to trace the Boolean operation sequence so that we can duplicate object with same shape but smaller volume. In order to get all the nodal information, the algorithm needs to traverse down to the primitive nodes of the CSG tree.

For the above reasons, the program is implemented in C language that allows recursive function calling and tree data structure handling. A function Solid_Offset() will call itself repeatedly until it reaches the terminating condition. The terminating condition is the recognition of the primitive node where dimension parameter values will be offset positively or negatively. Control is then passed back to the parent node and the process repeats until the whole CSG tree is traversed. The flow chart for the recursive function calling control is shown in Fig. 11.

Notes:

I. TP/C = Test primitive/composite and pass control to search for the 1st child node.

II. P/R = If the child is primitive (P) then change its dimension parameters and return to the next child node.

III. RN = Return control back to the nearest parent node and pass the control to 2nd child subtree.

There are two core routines in the implementation for void making, i.e. generating thin shell solids.

2.6.2.1. *Function Solid_Offset()*

This is the recursive function that will analyse the ultimate solid S. This function first checks whether the input solid is a primitive or a composite. If it is a primitive, the solid id will be passed to another function, Primitive_Offset(), to alter its parameters. If the id belongs to a composite, the program needs to classify which Boolean operation (Union, Intersection or Subtraction) is used for the combination. Under each Boolean operation case, there is an executable statement that calls Solid_Offset() again. The CSG tree can thus be traversed recursively

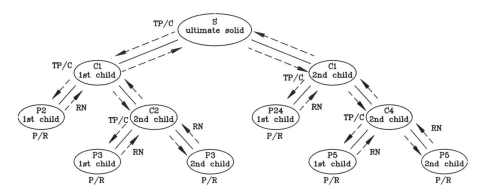

Fig. 11.   Flow chart for recursive function calling control.

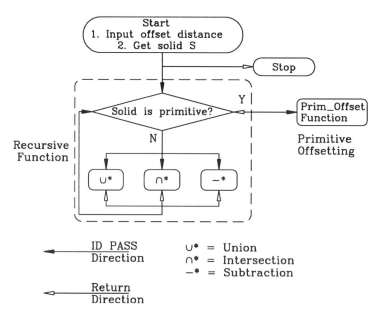

Fig. 12.   Flow chart for solid offset.

until all primitives are processed. The flow chart for offsetting the solid is depicted in Fig. 12.

The algorithm for the Solid_Offset() function is

Procedure *Solid_Offset(id, offset)*
/* id stores an identifier, offset stores the offset distance */
{ **switch** (*id*) {
  case *PRIMITIVE:*
        return(Primitive_Offset(id, offset));
        break;
  case *UNION:*
        return(Union_Solid(Solid_Offset(1st_child_node, offset),
                          Solid_Offset(2nd_child_node, offset)));
        break;
  case *INTERSECTION:*
        return(Intersection_Solid(Solid_Offset(1st_child_node, offset),
                            Solid_Offset(2nd_child_node, offset)));
        break;
  case *SUBTRACTION:*
        return(Subtraction_Solid(Solid_Offset(1st_child_node, offset),
                            Solid_Offset(2nd_child_node, offset)));
        break;
} **End Switch**
} **End Procedure** *Solid_Offset*

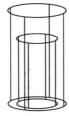

Fig. 13.   Bias offset of box, wedge, cylinder and cone.

Table 1.   Compensation offset matrix.

Primitives	Offset matrix (x, y, z)
Block	(-offset, -offset, -offset)
Wedge	(-offset, -offset, -offset)
Cylinder	(0, 0, -offset)
Cone	(0, 0, -offset)
Sphere	not required
Torus	not required

2.6.2.2. *Function Primitive_Offset()*

Here, we need to consider the offset matrix for each type of primitive. Suppose a sphere is being negatively offset, then a smaller diameter sphere is created. This sphere will also inherit the position matrix of the original sphere, i.e. the centres of the original sphere and the offset sphere are coincided. The same is applicable to the torus case but not to cases of block, wedge, cylinder and cone if the local coordinate system origins for block, wedge, cylinder and cone primitives do not coincide with their centroids. (See Fig. 13.) Primitive type dependent offset matrix is therefore employed to avoid offsetting about the origins of the local coordinate system. (See Table 1.) The new position matrix for the offset primitive is obtained by multiplying the original position matrix with the offset matrix.

In addition, uniform offset, i.e. constant offset thickness from all boundary surfaces, is preferred. Therefore, inclined surfaces of wedge and cone primitives are compensated by using formulae as illustrated in Fig. 14.

2.7. *Practical limitations*

In practice, the offset theory cannot be directly applied to produce thin shell rapid prototype. Some practical issues relating to the problem of building internal shell with no interior support should be taken care of.

(i) Sag of large top surface

This problem is closely related to liquid resin prototyping whilst the prototype is created by curing the liquid resin with laser. The liquid being solidified is, however,

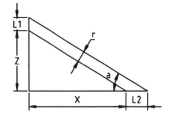

r = Offset distance

L1 = r/cos(a), L2 = r/sin(a), tan(a) = Z/X

$$\text{Z direction offset value} = L1 + r = r\left[1 + 1/\cos(a)\right] = r\left[1 + \text{sqrt}(X^2 + Z^2)/X\right]$$

$$\text{X direction offset value} = L2 + r = r\left[1 + 1/\sin(a)\right] = r\left[1 + \text{sqrt}(X^2 + Z^2)/Z\right]$$

Fig. 14.   Uniform offset compensation for inclined surface.

still in a soft state. A large soft top surface will deflect due to its own weight and the surface will no longer be flat. (See Fig. 15(a).)

(ii) Drop of submerged portion

Since the prototyping process is built from the bottom up layer by layer, the bottom surface of the submerged region that has no support underneath will definitely drop if the overhang distance is too large. (See Fig. 15(b).)

(iii) Drop of internal shell

An enclosed cavity in the original solid will become a free floating shell after the offset. In practice, a free floating shell without any interior support cannot be produced. (See Fig. 15(c).)

Although most of the commercially available RP system (such as QuickCast[25]) can generate interior support automatically, the generated interior support pattern is usually a grid pattern of regular size. Adaptive methodology should be used to generate the interior support pattern. Adaptive support can make the optimum use of the overhang capability of the prototyping material and be able to generate variably sized support pattern according to the dimensions of the various features in the solid model. The above problems can be solved by the sub-boundary octree approach. The approach will produce a thin shell rapid prototype with reinforced interior structure.

3. Octree Reinforced Thin Shell Solid

A reduced volume solid from negative solid offset requires interior reinforcement structure to solve the manufacturing problems. The reinforced hollow solid can be produced with the help of a sub-boundary octree. A sub-boundary octree is an octree[26] that is wholly in the interior of the solid. In this section, the background

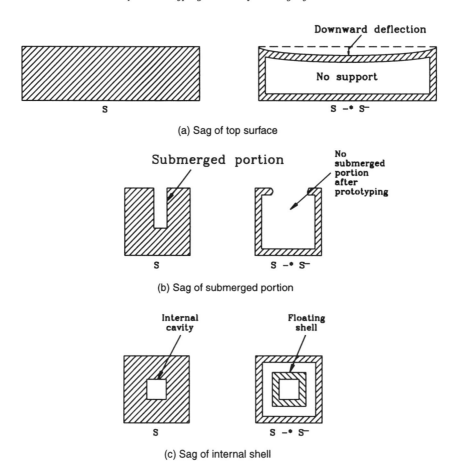

Fig. 15. Problem of building internal shell without interior support.

theory and implementation for generating octree reinforced thin shell solids are explained. Solid models defined by both CSG and STL representations are investigated. By using the CSG approach, the reinforced hollow solid can be obtained by:

$$RVS = S -^* S^{\text{octant}}. \tag{1}$$

In Eq. (1), S^{octant} is created from the sub-boundary octree. (To be discussed later.) By differencing S^{octant} from S, a thin shell hollow solid with interior reinforcement structure can be produced. Another alternative approach is based on STL boundary represented solid model. In this case, the reinforced hollow solid is determined by the following equation:[27]

$$RVS = (S -^* S^-) \cup^* S^{\text{octree}}. \tag{2}$$

Simply speaking, $(S -^* S^-)$ represents the thin shell hollow solid while S^{octree} represents the octree reinforced solid framework from S^-. The solid octree reinforcement, S^{octree}, is unioned with the thin shell hollow solid to solve the problem of providing necessary interior supporting structure.

3.1. *Implementation*

3.1.1. *Sub-boundary octree extraction*

The sub-boundary octree is extracted by classifying each octant as inside, outside, or partial after each subdivision of the solid object.[26] If the octant is classified as inside the solid object, the octant is stored in a list of sub-boundary octree. Further subdivision of intersecting octant is continued and the subdivision is terminated at either one of the following two conditions: (a) the user specified maximum level of subdivision is reached, and (b) there is no intersection between all the subdivided octants and the solid object. The initialization algorithm for subdivision and the recursive subdivision algorithm are presented in Ref. 28 and summarized by the following pseudo code:

Procedure *StartSubdivide*
/* *Initialization procedure of the subdivision process* */
{   *Cube ← Find the bounding cube of the object;*
       Subdivide (Cube);
} **End Procedure** *StartSubdivide*

Procedure *Subdivide (Cube_to_subdivide)*
/* *octants[8] store the eight subdivided octants by subdividing Cube_to_subdivide* */
{   **If** (*level of subdivision > maximum level of subdivision*)
       then *return;*
       octants[8] ← Subdivide Cube_to_subdivide into eight octants;
       While (*Not all octants are classified*)
       {   *octant ← Take one octant from octants[8];*
              If (*octant is classified as PARTIAL*)
              then
              {   *octant_to_subdivide ← octant;*
                     Subdivide (octant_to_subdivide);
              }
              else
              {   /* *Check whether the octant is inside the object* */
                     If (*octant is classified as INSIDE*)
                     then
                            Store_in_sub_boundary_octree (octant);
                     End If
              } **End If**
       } **End While**
} **End Procedure** *Subdivide*

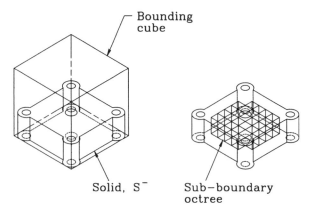

Fig. 16.   Solid, bounding cube and sub-boundary octree.

A sample solid object, its bounding cube and the generated sub-boundary octree with a maximum subdivision level of 3 are shown in Fig. 16. As the subdivision algorithm is a recursive procedure, the program is implemented in C++ language that allows recursive function calling.

The main problem in sub-boundary octree extraction is to classify the octants. For CSG solids, which is one of the popular solid representation schemes, efficient octant classification algorithm is reported.[29] Other than CSG, another popular solid representation scheme is boundary representation. In today's RP industry, the STL format is the de facto standard for product data exchange, especially with CAD. In essence, the STL data are triangles data set obtained from the tessellation of the solid boundary. However, known algorithm for octant classification of solids represented in STL format is not available. A simple and efficient octant classification algorithm for solids represented in STL format is thus described first in the following section.

3.1.1.1. Octant classification for solids represented by STL

With the STL format, solid is represented by a set of tessellated triangles. In the STL data file, the first 84 bytes are used to provide the header information. Each tessellated triangle is then described by its three vertices and surface normal. As a floating-point number in the STL file is represented by 4 bytes, the geometric description data of each tessellated triangle occupies 48 bytes (i.e. 4*4*3). After reading the entire STL file of the solid, the bounding box of the solid object can be determined by sorting the x-, y- and z-coordinate of the vertices of all triangles. The bounding box is then used to eliminate all trivial outside octants, which do not intersect the bounding box. Since the solid is represented by a set of triangles, the classification problem is reduced to the determination of intersections between the octants and the triangles. Thus a bounding box test of a triangle is employed to speed up the classification algorithm. The algorithm for classifying an octant against solid object represented by STL is presented as follows:

Procedure *ClassifyOctantVsSolid (Oi, S)*
/*   *Oi is the octant to be classified against the solid object S. The result to be*
          returned is either IN, OUT, NIO (i.e. inside, outside, or partial).
*/
{   **If** *(Oi does not intersect with the bounding box of S)*
         then *return OUT;*
         End If
         While *(Not all tessellated triangles are classified)*
         {   *Ti ← Take one tessellated triangle of ∂S;*
                  If *(Oi intersects with the bounding box of Ti)*
                  then
                  {   **If** *(Oi intersects with Ti)*
                           then *return NIO;*
                           End If
                  } **End If**
         } **End While**
         /* *Up to this point, Oi is either IN or OUT* */
         return CheckOctantINorOUT (Oi, S);
} **End Procedure** *ClassifyOctantVsSolid*

The procedure CheckOctantINorOUT is used to determine whether an octant is inside or outside a solid by projecting a ray, say, in the positive y-direction from a point, say P, on the diagonal of the octant. (See Fig. 17.) If the number of intersections between the ray and the tessellated triangles is an odd number, the octant is classified as inside, i.e. IN; otherwise an OUT is returned.[26]

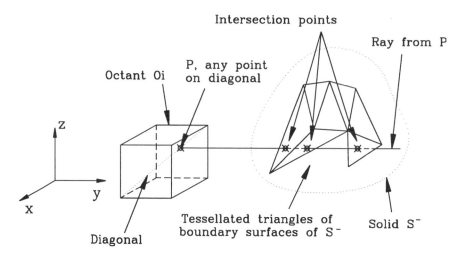

Fig. 17.   Octant and its projection ray.

3.1.2. *Preliminary check*

In order to verify the correctness of the algorithms for both the CSG and STL approaches, some tests are being done on simple solids. These include sphere, cylinder, and wedge. The resulting sub-boundary octree for a maximum subdivision level of 2 are shown in Fig. 18. Note that all the simple solids being used in the preliminary tests are solids with zero genus. A sample solid object with genus equal to five (Fig. 32) is used in the following sections to demonstrate the applicability of the algorithms to more complex solid.

3.1.3. *Creating octree reinforced hollow solid for the CSG approach*

The procedure for generating the reinforced hollow solid in the CSG approach is shown in Fig. 19. After extracting the sub-boundary octree, a list of 'INSIDE' octants is obtained. Firstly, a core is generated from each 'INSIDE' octant. The core is defined as the union of three rectangular blocks in the x-, y-, and z-direction respectively. (See Fig. 20.) The centres of the three rectangular blocks are the same

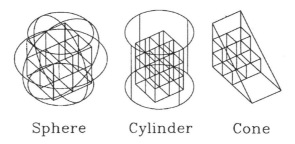

Sphere   Cylinder   Cone

Fig. 18.   Solid and their sub-boundary octrees in preliminary check.

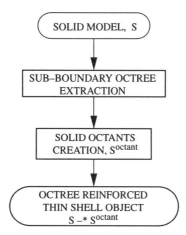

Fig. 19.   Generating octree reinforced RVS by CSG approach.

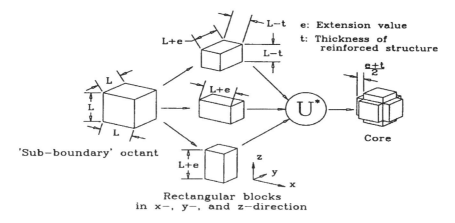

Fig. 20. Primitive core for CSG approach.

as the centre of the 'INSIDE' octant. Since the Boolean union operation is not stable for touching solids, especially for a large number of Boolean operations, the longest dimension of the three rectangular blocks are set to the side length of the cube plus a small extension value. The small extension is used to make sure that adjacent cores will penetrate into one another. The side length of the square section of the rectangular blocks is equal to the side length of the octant minus the user-specified thickness of the reinforced structure.

By using the above method, a pattern of cores is created from the extracted sub-boundary octree. (See Fig. 21.) By unioning all the cores, S^{octant} is obtained. By using Eq. (1), the reduced volume solid can then be generated by differencing S^{octant} from S. In Fig. 22(a) and (b), the reduced volume solid is sectioned to show its internal structure. For the test solid shown in Fig. 21, 56 cores, i.e. 56×3 or 168 rectangular blocks, are created from the 56 sub-boundary octants. This also involves $56 \times 3 - 1$ or 167 union operations.

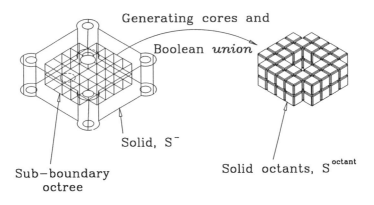

Fig. 21. Solid octants generation.

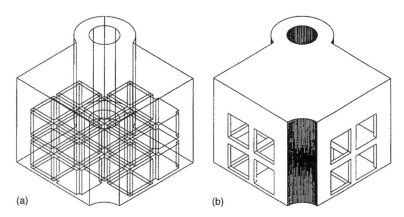

Fig. 22. Sectioned view of RVS generated by CSG approach (a) Wireframe model; (b) Hidden lines removed model.

As extension of the longest dimension of the rectangular blocks is required to avoid the touching numerical instability and the number of solid rectangular blocks for generating the solid octant cores is 3 times the number of 'sub-boundary' octants, the CSG approach is not very efficient. For the test solid in Fig. 22, the thickness of the reinforced structure is 3 mm. The volumes of the original test solid and its reduced volume solid are 409,350 mm^3 and 265,929 mm^3 respectively. Thus, a volume reduction of 35% is produced from the 56 cores.

3.1.4. *Creating octree reinforced hollow solid for the STL approach*

The procedure for generating the reinforced hollow solid from STL represented solid is shown in Fig. 23. Note that a "skeleton extension" step is required to avoid the problem of a floating skeleton. (See Fig. 24.) The sub-boundary octree solid, i.e. S^{octree}, can then be created by fleshing out the extended skeleton. This is done by adding width and thickness to the skeleton.

3.1.4.1. Skeleton extraction algorithm

By using the sub-boundary octree extraction algorithm mentioned in Sec. 3.1.1, the octants belonging to the sub-boundary octree are passed to an octree skeleton extraction algorithm for extracting and updating the skeleton framework. As the octants are positioned with their bounding edges lying in the principal directions of the coordinate system of the solid object, the bounding edges of the octants are categorized into x_edge, y_edge and z_edge for edges lying in the x-, y- and z-direction respectively. The skeleton extraction algorithm is presented in the following:

Procedure *Skeleton_extraction (octant)*
/*   *x_edge[4], y_edge[4] and z_edge[4] are used to store the twelve boundary edges of the octant. Both the edge starting point and its length will be modified as appropriate.*
*/

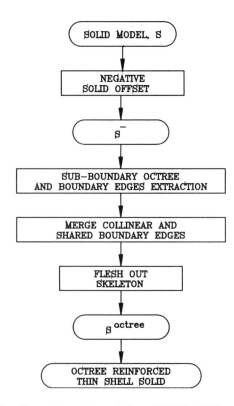

Fig. 23.   Generating octree reinforced RVS by STL approach.

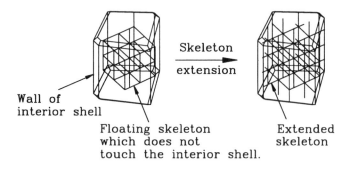

Fig. 24.   Floating skeleton and skeleton extension.

```
{   x_edge[4] ← Boundary edges of octant in x-direction;
    y_edge[4] ← Boundary edges of octant in y-direction;
    z_edge[4] ← Boundary edges of octant in z-direction;
    Extract_and_merge_x_skeleton ( x_edge[4] );
    Extract_and_merge_y_skeleton ( y_edge[4] );
    Extract_and_merge_z_skeleton ( z_edge[4] );
} End Procedure Skeleton_extraction
```

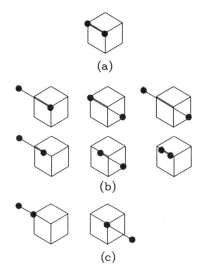

Fig. 25.   Edges that are (a) wholly overlapped; (b) partially overlapped; (c) touching.

In order to reduce the amount of union operations during the construction of sub-boundary octree solid from the extracted skeleton edges, edges that are wholly overlapped, partially overlapped or touched at one end should be merged to form a single edge. (See Fig. 25.) During the merge, the starting point and length of the tested skeleton edge is modified according to the overlapping or touching condition. (See Fig. 26.)

One advantage of the STL-based approach is that collinear boundary edges of adjacent octants can be merged together before being fleshed out to create the solid skeleton. In other words, 2D union of the boundary edges of the sub-boundary octants can be performed instead of 3D union of the solid skeletons. (See Fig. 27.) As a result, the number of solid skeletons and the required Boolean union operations for generating S^{octree} from the solid skeletons is much less than their CSG counterparts. In addition, the touching face problem for adjacent cores in the CSG approach does not exist in the STL-based approach. This is because collinear touching edges are merged at the edge level, i.e. the 2D level, and not at the solid level.

By using Eq. (2), the reduced volume solid can be generated. In Fig. 28(a) and (b), the generated reduced volume solid is sectioned to show its internal reinforced structure. For the test solid shown in Fig. 28, 80 solid skeletons are created from the sub-boundary octree. This is much less than the 168 rectangular blocks required in the CSG approach. Thus, there is over 50% reduction in the number of solids to be unioned. With a 3 mm thick reinforced structure, the volumes of the original test solid and its reduced volume solid obtained by STL approach are 409,350 mm^3 and 218,389 mm^3 respectively. A volume reduction of 46.7% is achieved. Considering the 35% volume reduction achieved by CSG approach, an 11.7% improvement in volume and material reduction is obtained. The higher volume reduction is due to the fact that a thin shell solid with uniform thickness is generated from $(S -^* S^-)$

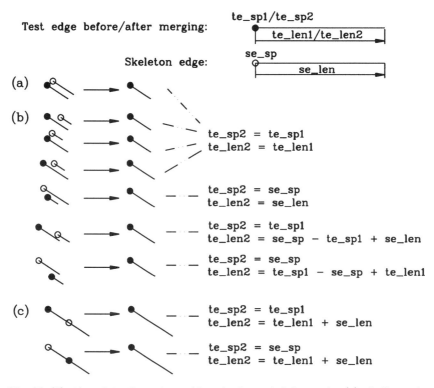

Fig. 26.   Modification of starting point and length of tested skeleton edge (a) wholly overlapped; (b) partially overlapped; (c) touching.

as shown in Eq. (2) and Fig. 28(b). On the contrary, thin shell solid produced from Eq. (1) has a non-uniform thickness as shown in Fig. 22(b).

4. Thin Shell Rapid Prototyping by FDM — Case Studies

4.1. *Preliminary overhang study*

A testing model with overhang distance from 5 to 10 mm (in step of 0.5 mm) and 10 to 20 mm (in step of 2 mm) is built by FDM. This testing model is used to check the capability of the FDM system in producing the reinforced thin shell rapid prototype. The testing part is generated successfully although some of the filaments dropped off. (See Fig. 29.)

4.2. *Feasibility study on building reinforced thin shell rapid prototype*

In order to check the capability of the FDM system in producing octree reinforced thin shell rapid prototype, a testing model which resembles the reinforced octree structure framework is used (Fig. 30). As shown in Fig. 31, some of the filaments

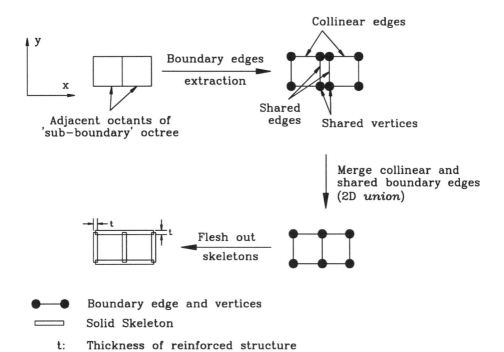

Fig. 27.   Boundary edges processing and fleshing out solid skeletons.

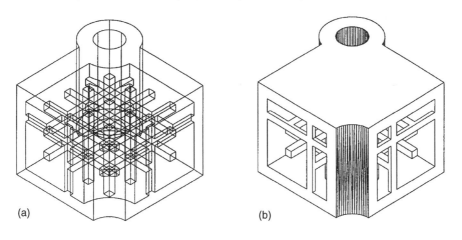

Fig. 28.   Sectioned view of RVS generated by STL approach (a) Wireframe model; (b) Hidden lines removed model.

Fig. 29.   Overhang testing model for FDM.

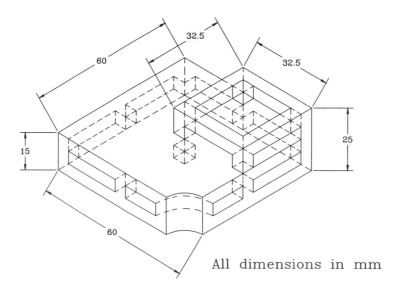

Fig. 30. Testing model with reinforced octree framework.

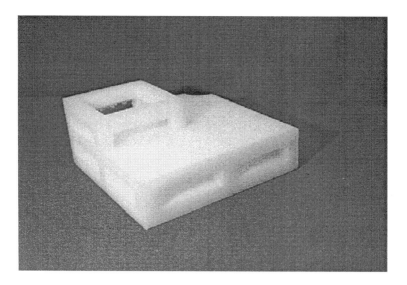

Fig. 31. Successful testing model with reinforced octree framework.

dropped off in building the first layer of the overhang ceiling. However, the second layer of the overhang ceiling can already obtain enough support from the previous layer. After four layers, the material filling becomes normal and no drop off is observed. The range for the dimensional accuracy of the successfully constructed testing model is within 0.00 to −0.24 mm, as shown in Table 2. Since the dimensional accuracy of the testing model from FDM is good and the model is constructed without any defect, the FDM is capable to produce the octree reinforced thin shell model.

Table 2. Dimensional accuracy of built testing model.

Dimension to be checked	Design dimension (mm)	Measured dimension (mm)	Deviation (mm)
1st step length	60.00	59.82	0.18
1st step width	60.00	59.78	0.22
1st step height	15.00	14.84	0.16
2nd step length	32.50	32.36	0.14
2nd step width	32.50	32.50	0
2nd step height	25.00	24.76	0.24

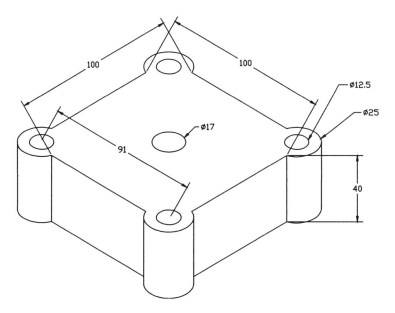

Fig. 32. Solid model of test part.

4.3. *Reinforced thin shell rapid prototype of 2.5 mm wall thickness*

Figure 32 shows the thin shell rapid prototype of a solid model being used to validate the proposed octree reinforcement technique. As it is hypothesized that the wall thickness of the thin shell must be within 3 to 5% of the longest side of the solid model, a testing model with a wall thickness of 2.5 mm and reinforcement support of $3 \times 3 \, \text{mm}^2$ is created. The partial section of the constructed prototype is generated to investigate the internal details of the reinforced structure, as shown in Fig. 33. From the test, it is observed that some filaments dropped off in filling the first two layers of the top ceiling. However, the joining of the filaments returns to normal when the third layer is drawn. When the eighth layer is drawn, no significant defect in the top ceiling is observed. Since the layer thickness of the FDM is 0.254 mm, the minimum layer thickness of the thin shell model should be greater than 2.032 mm

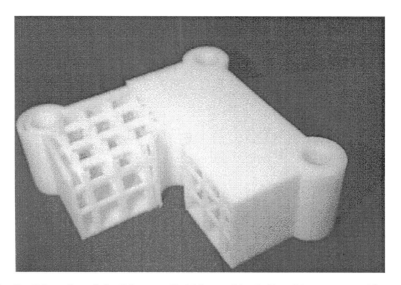

Fig. 33.   Partial section of the 2.5 mm wall thickness thin shell rapid prototype with 3×3 mm^2 reinforced structure.

Table 3.   Comparison of volume, time and cost for producing reinforced thin shell rapid prototype and solid prototype.

	Solid prototype	Reinforced thin shell prototype
Material volume (mm^3)	409,350	144,422
Production time (hr)	46.9	26.5
Manufacturing cost (HK$)	5,344	2,883
Volume reduction (%)	N/A	64.72
Time saving (%)	N/A	43.5
Cost saving (%)	N/A	46.05

(i.e. 8×0.254 mm). Table 3 shows the comparison of material volume, production time and manufacturing cost for producing the thin shell prototype and the solid prototype.

5. Discussions and Future Works

In practice, the prototypes produced by FDM will contract when the material is cooled down. The reinforced thin shell prototypes are thus examined to see whether there is any significant change in dimensional accuracy and surface flatness. By measuring the reinforced thin shell prototype, it is found that the range of deviation from the design dimension, as shown in Fig. 32, is from 0.16 to 0.22 mm, with an average deviation of 0.18 mm. From the measurement, we can see that the dimensional accuracy of the reinforced thin shell prototypes are good and there is no evidence to show that the dimensions are changed drastically. With dial gauge checking, it is found that the maximum deflection of the top surface of the prototype is 0.072 mm.

The average deflection is 0.052 mm. The standard deviation of the deflection is 0.017 mm. Thus, there is no significant variation in surface flatness of the reinforced thin shell prototype. From Table 3, we can see that the proposed approach is quite successful in terms of the 65% material saving, 44% faster production time and 46% cheaper manufacturing cost.

In addition, the sub-boundary octree solid should be oversized to penetrate into the inner wall of the thin shell solid. This is to overcome any numerical instability in performing large amount of union operations due to the touching faces between the inner wall of thin shell and the sub-boundary octree solid.

One advantage of reinforced thin shell rapid prototyping technique is in rapid tooling generation. The thin shell prototype will serve as a core and enclosed by a ceramic shell. The core will then be burnt to leave a cavity. A solid prototype will, however, expand and break the shell. The thin shell prototype will not have this problem as the prototype will break inward. The thin shell prototype can be tested by castable ceramics rapid tooling for investment casting operations. The thin shell pattern and the completed ceramic shell assembly will be placed in a fully aspirated oven and fired at over 980°C for one hour. After the burnout is completed, any small cracks or fractures occurred in the investment shell will then be inspected.

Another advantage of the proposed technique is in the strength reinforcement of the thin shell prototype. The strength of the prototype can be determined experimentally by, say, non-destructive optical method. The experimentally obtained strength can be compared with the strength of the solid object to determine the effectiveness of the prototype. Numerical model, such as finite element model, can also be developed to predict the strength of the reinforced prototype.

Notations

$\forall$: universal quantifier

$\in$: belongs to

$\cap, \cap^*, \cup, \cup^*, -, -^*$: non-regularized and regularized intersection, union and difference

$\subset$: proper sub-set of

$\subseteq$: improper sub-set of

$\emptyset$: null set

U: universal set (Euclidean space) $= cA \cup \partial A \cup iA$

A, B, S: regularized sets, i.e. solid volumes

RVS: reduced volume solid, i.e. thin shell solid

S^{octree}: sub-boundary octree

S^{octant}: solid octants of the sub-boundary octree

∂A: boundary of A

iA: interior of A. $A = \partial A \cup iA$. $\partial A, iA \subset A$ if $A \neq \emptyset$

cA: complement of A

c^*A: regularized complement of A

A^+: positive offset by distance r, $A^+ = \{x : \exists y \in A \text{ such that } \|y - x\| \leq r\}^1$
A^-: negative regularized offset by distance r, $A^- = c^*((c^*A)^+)^1$

References

1. J. R. Rossignac, Blending and offsetting solid models, PhD Thesis, University of Rochester, 1985.
2. P. Jacobs, Rapid prototyping & manufacturing: Fundamentals of stereolithography, Society of Manufacturing Engineers, 1992.
3. B. Haase, Testing rapid prototypers, *MicroCAD News*, October 1991, 27–29.
4. S. W. Thomas, Stereolithography simplifies tooling for reinforced rubber parts, *Mechanical Engineering*, July 1992, 62–66.
5. K. M. Gettelman, Stereolithography: Fast model making, *Modern Machine Shop* **62**, 5 (October 1989) 100–107.
6. C. R. Deckard, *Selective laser sintering (CAD/CAM)*, PhD Thesis, University of Texas at Austin, 1988.
7. Stratasys, Inc., Fused deposition modeling for fast, safe plastic models, *NCGA '91 12th Annual Conference & Exposition*, Chicago, IL, USA (22–25 April 1991) 326–332.
8. M. Feygin, *et al.*, Laminated object manufacturing (LOM): A new tool in the CIM world, *IFIP Transactions B* **B-3** (1992) 457–464.
9. http://www.sanders-prototype.com/main.html.
10. http://www.zcorp.com/frame_build.html.
11. 3D Systems, Inc., *Stereolithography Interface Specification*, part number 50065-S01-00, October 1989.
12. National Bureau of Standards, *Initial Graphics Exchange Specification* Version 4.0, 1988.
13. B. Miller, Fast prototyping makes models sooner and better, *Plastics World*, February 1991, 44–47.
14. S. J. Muraski, Make it in a minute, *Machine Design*, 8 February, 1990, 127–132.
15. I. P. Faux and M. J. Pratt, *Computational Geometry for Design and Manufacture*, Ellis Horwood, 1981.
16. R. S. Pressman and J. E. Williams, *Numerical Control and Computer-Aided Manufacturing* (John Wiley & Sons, 1977).
17. B. Pham, Offset curves and surfaces: A brief survey, *Computer Aided Design*, April 1992, 223–229.
18. S. E. O. Saeed, A. de Pennington and J. R. Dodsworth, Offsetting in geometric modelling, *Computer Aided Design*, March 1988, 67–74.
19. R. H. Wang and W. H. Jiang, Algorithm of the offset shape, *Computers & Graphics*, 1991, 435–439.
20. W. R. S. Sutherland, The offsets of a convex polygon, *Methods of Operations Research* **62** (1990) 33–41.
21. W. Tiller and E. G. Hanson, Offsets of two-dimensional profiles, *IEEE Computer Graphics & Applications*, September 1984, 36–46.
22. Y. J. Chen and B. Ravani, Offset surface generation and contouring in computer-aided design, *Journal of Mechanisms, Transmissions, and Automation in Design* **109**, 1 (March 1987) 133–142.
23. R. T. Farouki, Exact offset procedures for simple solids, *Computer Aided Geometry Design* **2**, 4 (December 1985) 257–279.
24. S. Aomura and T. Uehara, Self-intersection of an offset surface, *Computer Aided Design*, September 1990, 417–422.

25. K. W. Lee, *Principles of CAD/CAM/CAE Systems* (Addison Wesley, 1999).
26. I. Zeid, *CAD/CAM Theory and Practice* (McGraw-Hill, 1991).
27. K. M. Yu, T. W. Lam, K. M. Cheung and C. L. Li, Thin-shell based rapid prototyping, *Proceedings of 3rd International Conference on Manufacturing Technology*, Hong Kong, 1995, 230–233.
28. T. W. Lam, K. M. Yu, K. M. Cheung and C. L. Li, Octree reinforced thin shell objects rapid prototyping by fused deposition modeling, *International Journal of Advanced Manufacturing Technology* **14**, 9 (1998) 631–636.
29. Y. T. Lee, *et al.*, Algorithms for computing the volume and other integral properties of solids, *Communications ACM* **25**, 9 (1982) 635–650.

CHAPTER 3

COMPUTER TECHNIQUES APPLICATIONS IN OPTIMAL DIE DESIGN FOR MANUFACTURING SYSTEM

JUI-CHENG LIN

National Hu-Wei Institute of Technology,
Department of Mechanical Design Engineering,
Taiwan 63208, R.O.C.

Die casting processes are widely prevalent in manufacturing systems. They involve many complex factors that determine casting quality and production. This chapter provides an in-depth treatment of this subject and points out that die cast testing time and costs can be reduced by the methods presented. Further, new and highly effective techniques are introduced to predict die casting performances very accurately.

Keywords: Optimal die design; die casting processes; finite element analysis.

1. Introduction

The manufacture of die-casting molds or injection molds is a precise yet clean production process. Casting is applied widely in motor, airplanes, ship, electrons, precise machinery etc, because it can produce higher strength high quality casting. It has a profound influence on the speed and accuracy demanded by modern industries. Die casting is a precise casting method of inject molten metal into die cavity giving high pressure (cold chamber method: 170–2000 km/cm^2; hot chamber method: 90–500 km/cm^2) and making use of high velocity (20–60 m/s). Whether the forming of casting is successful or not is always determined by the casting flow system. The die-casting processes are many complex factors that determine casting quality and production. During the performance in the die casting process, the plane of the casting generally indicates the accuracy of the mold-surface profile. The arrangement and shape of the gate, runner, sprue, pouring basin, overflow, air vent etc, are the most important factors in the design of the die-casting die. Mistakes design of the die casting die would influence the accuracy or produce defects in the part. The shape of a die-casting part could be complex in a practical case, so the factors will increase the influence to the casting part.

Generally speaking, die design still depends on experience in handing over, due to the lack in analytic ability in die and melting metal flow and heat transfer. Die design is unable to know and handle the deformation resulting from material and thermal expansion and shrinkage of the die. The major cause is injection from different position (finding of die plane-line), the flow cause tremendous temperature distribution, and cause different shrinkage, thus resulting in poor die design.

Though finite elements analysis software is capable in analyzing the flow condition of injection metal and the stress, strain and temperature distribution condition of product (work piece) under various injection conditions, the establishment of analytic model is very difficult. Besides understanding the die, metal flow and solidification process, user should fully acquaint with the basic finite element software. Integration purpose can be achieved and a lot of money and time saved only if one completely understand the process of die manufacturing, and eliminate the annoyance cause by the moving of personnel.

Initially, consider the design of vent gate and overflow gate in the process of injection and flow to fill the die cavity during cold room die-casting denoted in the research of flowing simulation by Garber.[1] When the metal casting through plunger into cavity, consider change occurred in metal, and transform the air in the cavity into molten metal. Subsequently, Garber[2,3] studied too large or too small plunger speed that will affect the cast quality. Groenevelt et al.[4] studied and discussed the influence from speed of injection molten metal into cavity, flow distance and cavity temperature to the quality of product after casting. According to the experiment, distance of molten metal flow increase linearly while the die-casting speed increase. So do the die temperature — the range of temperature is approximately from 121 to 288°C. Truclove[5] using cooling system to control the overall temperature of the die in order to gain optimal heat transmission characteristic, reduced the occurrence of hard point phenomena in cast piece, in order to improve the quality of the piece. Kaiser[6] and Draper[7] proposed the importance of initial temperature (pre-heat temperature) of the die, and pointed out that too low pre-heat temperature would tend to fail in filling up the cavity inside die by die-casting liquid, and result in formation failure. Higher temperature probably increases the cooling time and reduces productivity.

General engineer think that the most important point in casting process is the flow process inside cavity, and Jong et al.[8] was devoted in developing a mathematics equation to express the flow and solidification of molten metal during the press of die-casting, in order to analyze the condition of temperature and solidification strain of die-casting components in the cavity.

New product is developed for improvement of defect of cast piece and facilitate the purpose of speedy and accuracy, Shibata[9-12] used limited elements method to analyze and design the die. The result not only elevated the accuracy, but also increased the factors to be considered, and the pressure of die-casting, speed of molten liquid flow, viscosity, and mechanical nature of material changed accompanied with temperature and phase.

As stated above, integration of CAD/CAE/CAM into one system is essential for the development of speedy and low cost die, and to facilitate the accuracy in simulation, detail assessment should be carried out in the overall integrated structure. To exert the 100% function of the software especially, Altan *et al.*[13] expressly proposed a computer integrated system CAM and provided a complete program subject to die manufacturer. According to the description of Hurt,[14] his basic CAD/CAE structure is capable to transform the geometrical shape into CAM portion according to the mathematics equation of shape, or the existing external configuration of the object and in conjunction with the simulated status of CAE, Corbett[15] additionally integrated a complete system for die manufacturing by using CAD system.

The main purpose of this study is to design and find the optimized manufacture variables of the die casting die, because only the best manufacture variables produces the minimum temperature difference in the cavity and the minimum deformation of the part. This study uses CAD/CAE/CAM software to systemic the design process of die, in order to minimize the human negligence in die design. Using the finite elements and CAM software to analyze the condition of die-casting, piece deformation after casting under various parameters (injection-position, the stress of the die-casting die, injection variable such as such as high-speed plunger injection position, runner injection angle and runner section ration, and the workpiece accuracy). And an abductive network is used to model the die casting process using casting experimented data. It has been shown that prediction accuracy in networks is much higher than that in a network.[16] Abductive networks based on the abductive modeling technique are able to represent complex and uncertain relationships between input and output variables. Once the abductive network has constructed the relationships of input and the output die-casting variables, an appropriate optimization algorithm with a performance index is able to search the optimal casting parameters.

In this study, a sound optimization method of simulated annealing[17] is adopted. The simulated annealing algorithm is a simulation of the annealing process for minimizing the performance index. It has been successfully applied to the filtering in image processing,[18] VLSI layout generation,[19] discrete tolerance design,[20] wire electrical discharge machining,[21] etc.

2. Abductive Network

It is well known that reasoning from general principles and initial facts to new facts is called deductive reasoning. However, the reasoning in a real problem is very often uncertain. Therefore another class of reasoning called abductive networks is used for reasoning from general principles and initial facts to new facts under conditions of uncertainty.

Miller[22] observed that human behavior, limits the amount of information at a time, then summarizing the input data, then passes the summarized information to a higher reasoning level. In an abductive network, a complex system can be decomposed into smaller, simpler subsystems grouped into several layers using polynomial function nodes. These nodes evaluate the limited number of inputs by a polynomial function and generate an output to serve as an input to subsequent nodes of the next layer. These polynomial functional nodes are specified as follows:

2.1. *Normalizer*

A normalizer transforms the original input variables into a relatively common region.

$$a_1 = q_0 + q_1 x_1 \tag{2.1}$$

where a_1 is the normalized input, q_0, q_1 are the coefficients of the normalizer, and x_1 is the original input.

2.2. *White node*

White node consists of linear weighted sums of all the outputs of the previous layer.

$$b_1 = r_0 + r_1 y_1 + r_2 y_2 + r_3 y_3 + \cdots + r_n y_n \tag{2.2}$$

where $y_1, y_2, y_3, \ldots, y_n$ are the input of the previous layer, b_1 is the output of the node, and the $r_0, r_1, r_2, r_3, \ldots, r_n$ are the coefficients of the triple node.

2.3. *Singles, double, and triples node*

These names are based on the number of input variables. The algebraic form of each of these nodes is shown in the following equation:

$$\text{single: } c_1 = s_0 + s_1 z_1 + s_2 z_1^2 + s_3 z_1^3 \tag{2.3}$$

$$\text{double: } d_1 = t_0 + (t_1 n_1 + t_2 n_1^2 + t_3 n_1^3) + (t_4 n_2 + t_5 n_2^2 + t_6 n_2^3) + (t_7 n_1 n_2) \tag{2.4}$$

$$\begin{aligned}
\text{triple: } e_1 = {} & u_0 + (u_1 o_1 + u_2 o_1^2 + u_3 o_1^3) + (u_4 o_2 + u_5 o_2^2 + u_6 o_2^3) \\
& + (u_7 o_3 + u_8 o_3^2 + u_9 o_3^3) + u_{10} o_1 o_2 + u_{11} o_2 o_3 \\
& + u_{12} o_1 o_3 + u_{13} o_1 o_2 o_3
\end{aligned} \tag{2.5}$$

where $z_1, z_2, z_3, \ldots, z_n$, $n_1, n_2, n_3, \ldots, n_n$, $o_1, o_2, o_3, \ldots, o_n$ are the input of the previous layer, c_1, d_1 and e_1 are the output of the node, and the $s_0, s_1, s_2, s_3, \ldots, s_n$, $t_0, t_1, t_2, t_3, \ldots, t_n$, $u_0, u_1, u_2, u_3, \ldots, u_n$ are the coefficients of the single, double, and triple nodes.

These nodes are third-degree polynomial equations and doubles and triples have cross-terms, allowing interaction among the node-input variables.

2.4. *Unitizer*

On the other hand, a unitizer converts the output to a real output.

$$f_1 = v_0 + v_1 i_1 \tag{2.6}$$

where i_1 is the output of the network, f_1 is the real output, and v_0, v_1 are the coefficients of the unitizer.

They have to give it enough parameters to train the database if we build a complete abductive network. Then, a predicted square error (PSE) criterion is used to automatically determine an optimal structure.[23] The principle of the PSE criterion is to select as accurate but as un-complex a network as possible. The PSE is composed of two terms, that is:

$$PSE = FSE + KP \tag{2.7}$$

where FSE is the average square error of the network for fitting the training data and KP is the complex penalty of the network, expressed by the following equation:

$$KP = CPM \frac{2\sigma_P^2 K}{N} \tag{2.8}$$

where CPM is the complex penalty multiplier, KP is the number of coefficients in the network, N is the number of training data and σ_P^2 is a prior estimate of the model error variance.

3. Simulated the Annealing Algorithm

Metropolis[24] proposed a criterion used to simulate the cooling of a solid for reaching a new state of energy balance. Based on the criterion of Metropolis, an optimization algorithm called "simulated annealing" was developed by Kirkpatrick.[25]

In this paper, the simulated annealing Algorithm is used to search for the optimal process parameters for die-casting. Figure 3.1 shows the flow chart in the simulated annealing. First, an initial temperature Ts, a final temperature Te, and a set of initial process parameter vector Ox. An objective function obj is defined based on the die casting performance index. The objective function will been recalculated using the perturbed compensation parameters. If the new objective function becomes smaller, the perturbed process parameters are accepted as the new process parameters and the temperature will drop a little, that is

$$T_{i+1} = T_i C_T \tag{3.1}$$

where i is the index for the temperature decrement and the C_T is the decaying ratio for the temperature ($C_T < 1$).

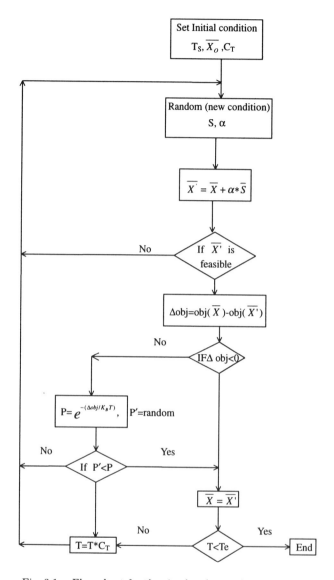

Fig. 3.1.   Flow chart for the simulated annealing searching.

However, if the objective function become larger, the probability of the acceptance of the perturbed process parameters is given as:

$$P_r(\text{obj}) = \exp\left[-\frac{\Delta\text{obj}}{k_B T}\right] \tag{3.2}$$

where k_B is the Boltzmann constant and Δobj is the difference in the objective
function. Repeat the above procedure until the temperature T approaches zero.
It shows the energy dropping to the lowest state.

4. The Runner Optimization Design of Die-Casting Dies

4.1. *The experimental setup*

In traditional die, the runner and body part (cavity block) are not separate. This study designs and makes the runner and cavity block separately, in order to establish the die casting results under different die casting conditions. That is, the twenty-two runner insert blocks have been place in the moving cavity block to conduct the die casting tests. Twenty-five experiments have been carried out. Figure 4.1 and Table 4.1 show the designs of the die and the runner blocks.

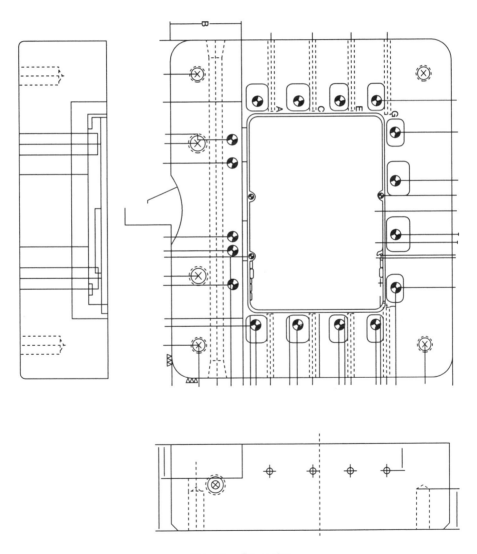

Fig. 4.1.   Die cavity.

Table 4.1. The design of die and runner blocks.

Type	Position	W	H	W/H	θ
1	90	18	9	2	60
2	90	18	6.5	2.8	60
3	90	18	9	2	80
4	90	18	9	2.8	80
5	110	18	6	2	60
6	110	18	10	2.8	60
7	110	18	7.5	2.8	80
8	80	18	6.5	2.4	70
9	120	18	6.5	2.4	70
10	110	18	6.5	2.4	45
11	100	18	6.5	2.4	90
12	100	18	10	1.8	70
13	100	18	6	3	70
14	100	18	6.5	2.4	70
15	100	18	6.5	2.4	70
16	100	18	6.5	2.4	70
17	100	18	6.5	2.4	70
18	100	18	6.5	2.4	70
19	100	18	6.5	2.4	70
20	100	18	5.5	2.4	70
21	100	18	6.5	2.4	70
22	100	18	6.5	2.4	70

First the cavity block of die casting die is installed into the die block, and then they are put in the die casting machine together. After that, set the condition of die casting experiment (e.g. casting pressure: $1200 \, \text{km/cm}^2$, low injection velocity: $0.2 \, \text{m/s}$, high injection velocity: $2.8 \, \text{m/s}$, cycling time of casting: 3 shots/min). After the molten aluminum has been injected into the die cavity, it solidifies because of losing heat. A die temperature too high or too low will have bad effects on the casting quality. It is very important to control the temperature of die casting but very difficult. First, one has to measure the temperature of die cavity surface, the position under the die cavity for temperature measurement, and one end of the thermal-couple was put at a point under the cavity of the moving die and the fixed die (points 1–16, as show in Table 4.2). The other end of the thermal-couple connects with the GPIB 556 temperature measuring equipment. As the operation of die-casting proceeds, start the GPIB 556 temperature equipment to measure the temperature of the points under the die cavity while the die cavity is opened and closed.

Table 4.2 shows the temperature distribution of a face ($145 \times 87 \, \text{mm}$) that is located under the cavity of the moving die and $20 \, \text{mm}$ away from the parting gate. It has a total of sixteen points. Based on the developed training database, a three-layer abductive network for predicting the distribution of the temperature of all area is automatically synthesized (Fig. 4.2). It utilizes this model to find the temperature of other points in the die-casting cavity (A–H). Figures 4.3–4.16 shows the distribution of temperature in the part of the 1–14th sets. They include the

Table 4.2. Measured point of temperature ($L * H = 145 * 87$ mm, total 16 point had observed temperature value (1–16), and 8 point unknown (A–H)).

	16	15	14	13			
1	E	F	G	H		12	
2						11	
3	A	B	C	D		10	
H 4						9	
	L→ 5	6	7	8			

Sets no.↓	1	2	3	4	5	6	7	8	9	10	11	12	13	14	15	16
1	135.5	146.6	149.4	144.3	187	177	182	186.9	164	170	155.7	156.4	155	150	145	140
2	143.9	151.8	153.3	147.2	191.3	181.3	186.3	191.3	163	170	164	162.9	160	155	152	145
3	128.1	136	145.7	138.9	175.2	165.2	170.2	175.3	158.6	160.6	150.9	150	150	140	136	130
4	138.2	149.3	152	138.4	185.3	175.3	180	185.3	157	162	153.6	153.6	154	152	150	142
5	147.1	156	161	153.3	200	190	195	200	170	177.2	170.8	170.3	165	160	160	161
6	157.5	168.5	172.9	162.5	205	195	200	204.5	184	186.6	178.4	178.8	178	171	165	162.4
7	146.1	156.5	157.4	148.4	198	188	193	197.5	172.7	179.7	167	167.7	165	159	157	147
8	143.4	151.2	157	148	199	189	193	198.4	173	175	169	166.4	163	157	155	145
9	157.2	166.3	171	164.5	209	199	203	208.8	178.8	185.2	178.6	176.6	174	171	168	171
10	153.5	162.6	166	157	208	198	201	207.5	186	191	174.9	174.6	172	166	162	156
11	148	156.5	162.3	156.4	187	177	190	187	161.9	162.2	161	160.2	160	158	156	150
12	147.2	157.6	162.5	158.1	185	175	185	184.8	160	166	160	158	158	155	155	150
13	131.2	140.7	146.2	140.8	190	180	190	190	168	168.7	157.7	156.3	156	150	145	135
14	158.2	166.9	170.6	160.6	210	200	210	209.8	184	186.7	179.4	176.9	176	171	171	161
15	155.4	156.6	171.3	162.4	208	198	208	207.6	184	189	179.4	171.1	174	169	167	159
16	158.9	169.9	170.3	160.1	208.3	198.3	208.3	208.2	183.3	189.7	180.7	180.4	177	174	171	161
17	155.2	166.1	166.7	155.2	203	193	203	202.6	177.6	184.2	175.6	175.4	173	168	164	158
18	153.7	165.5	170	159.4	204	194	204	203.7	179	188.5	176.4	174.9	174	169	164	158
19	157.2	169.9	174.1	161	207	197	207	207.3	182.3	189.3	179.6	178.5	177	171	165	159
20	157.4	167.1	170	157.2	207	197	207	206.8	176.9	186.6	177.6	177.5	176	171	165	159
21	155.5	164.1	167.3	155.4	205	195	204	204.9	180	185	175.6	174.9	173	169	164	159
22	153.2	161.7	166	156.4	206	196	206	205.5	175.5	183	174.6	173.3	171	170	168	158

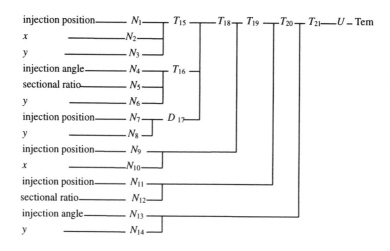

$N_1 = -9.59 + 0.0966x_1 \qquad N_2 = -1.29 + 0.118x_1 \qquad N_{3-} = -1.13 + 0.026x_1$

$N_4 = -6.22 + 0.0903x_1 \qquad N_5 = -6.7 + 2.76x_1 \qquad N_{6-} = -1.13 + 0.026x_1$

$N_7 = -9.59 + 0.0966x_1 \qquad N_{8-} = -1.13 + 0.026x_1 \qquad N_9 = -9.59 + 0.0966x_1$

$N_{10} = -1.29 + 0.118x_1 \qquad N_{11} = -9.59 + 0.0966x_1 \qquad N_{12} = -6.7 + 2.76x_1$

$N_{13} = -6.22 + 0.0903x_1 \qquad N_{14-} = -1.13 + 0.026x_1$

$T_{15} = 0.652 + 0.524x_1 + 0.238x_2 - 0.446x_3 + 0.0.096x_1^2 - 0.554x_2^2 - 0.183x_3^2 + 0.00428x_1x_2$
$\qquad\quad - 0.096x_1^3 + 0.0596x_2^3 - 0.12x_3^3$

$T_{16} = 0.174 - 0.278x_1 + 0.362x_2 - 0.446x_3 - 0.0752x_1^2 - 0.374x_2^2 + 0.248x_3^2 + 0.0508x_1x_2$
$\qquad\quad + 0.0269x_1x_3 + 0.0179x_1^3 - 0.181x_2^3 - 0.12x_3^3$

$D_{17} = -0.328 + 0.524x_1 - 0.446x_2 + 0.0.096x_1^2 + 0.248x_2^2 - 0.096x_1^3 + 0.0596x_2^3 - 0.12x_3^3$

$T_{18} = 0.195 + 0.909x_1 + 0.374x_2 - 0.47x_3 - 0.0568x_1^2 - 0.246x_2^2 - 1.16x_3^2 + 0.418x_1x_2$
$\qquad\quad + 1.28x_1x_3 + 0.118x_2x_3 + 0.369x_1x_2x_3 + 0.117x_1^3 + 0.316x_2^3 + 0.0838x_3^3$

$T_{19} = -0.578 + 1.16x_1 + 0.419x_2 + 0.277x_3 - 0.108x_1^2 - 0.443x_2^2 + 0.144x_3^2 - 0.025x_1x_2$
$\qquad\quad - 0.31x_1x_3 + 0.406x_1x_2x_3 - 0.0867x_1^3 - 0.219x_2^2 - 0.242x_3^3$

$T_{20} = 0.124 + 1.03x_1 + 0.0238x_1^2 - 0.0299x_2^2 - 0.107x_3^2 + 0.0387x_1x_3 + 0.533x_2x_3$
$\qquad\quad + 0.498x_1x_2x_3 - 0.0281x_1^3$

$T_{21} = 0.45 + 1.02x_1 + 0.0418x_3 + 0.0455x_1^2 - 0.0517x_2^2 - 0.069x_1x_3 - 0.0216x_2^3$

$U = 167 + 18.5x_1$

Fig. 4.2.   The abductive networks for predicting the distribution of the temperature of all area.

temperature of the A–H points. The average error is about 10% to the 15th–22th test data sets using this model. It shows that the models fit the experiments.

4.2. Finite element method

A 2-D rectangular section die cavity is shown in Fig. 4.1. We get the result from the finite element method; the software is COSMOS/M (version 2.6) and the pre- and

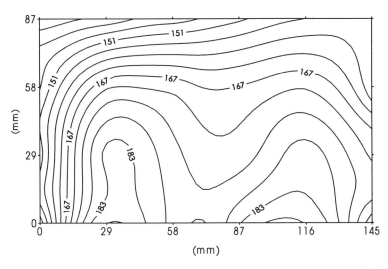

Fig. 4.3. Temperature distribution (no. 1 injection position = 90, runner sectional ratio = 2, runner injection angle = 60).

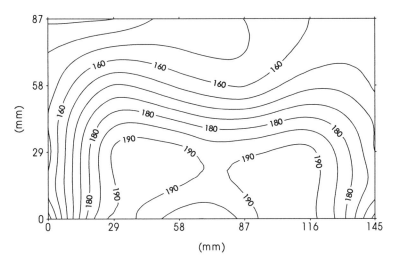

Fig. 4.4. Temperature distribution (no. 2 injection position = 90, runner sectional ratio = 2.8, runner injection angle = 60).

post-processor GEOSTAR of the Structural Research and Analysis Corporation (SRAC). Utilizing the NONLINEAR model analyses the change of the heat-strain in the THERMAL-LOADS condition. The plane is divided into 12 elements and twenty-four nodes. The initial values of temperature are shown in Table 4.2 (point 1–16) and point *A–H*; the boundary conditions are the nodes, five and eight, which are fixed. They are the metal-flow gates. The mechanical property of experimental material shows in Tables 4.3 and 4.4, the coefficient of thermal expansion and

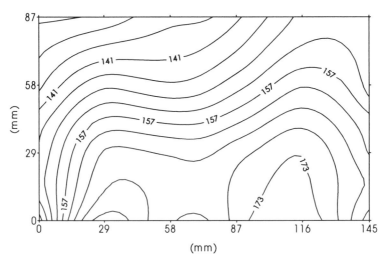

Fig. 4.5. Temperature distribution (no. 3 injection position = 90, runner sectional ratio = 2, runner injection angle = 80).

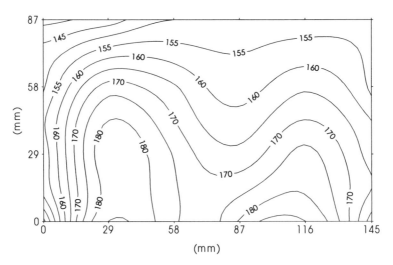

Fig. 4.6. Temperature distribution (no. 4 injection position = 90, runner sectional ratio = 2.8, runner injection angle = 80).

thermal conductivity depend on the temperature. Table 4.5 shows every part of the maximum component deformation. And Figs. 4.17–4.30 shows the deformation distribution for the 1th–14th sets. The 15th–22th sets are the same to check the repeatability on experiments.

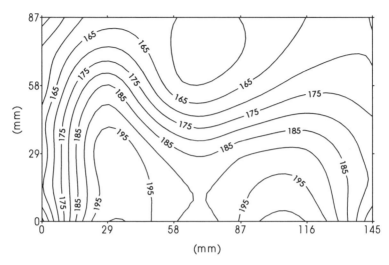

Fig. 4.7.   Temperature distribution (no. 5 injection position = 110, runner sectional ratio = 2, runner injection angle = 60).

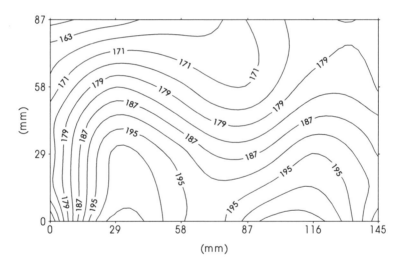

Fig. 4.8.   Temperature distribution (no. 6 injection position = 110, runner sectional ratio = 2.8, runner injection angle = 60).

4.3. *Creating the runner model*

Based on the developed training database, a three-layer abductive network for predicting the factor of design and the maximum deformation is automatically synthesized. It also accurately predicts the maximum deformation. All polynomial equations used in this network are listed in Fig. 4.31.

Table 4.6 shows the result of a comparison between the experimental maximum deformation and model predicted values, which are measured in a practical die

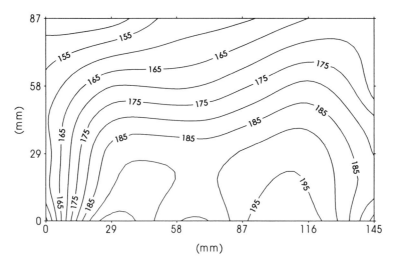

Fig. 4.9. Temperature distribution (no. 7 injection position = 110, runner sectional ratio = 2.8, runner injection angle = 80).

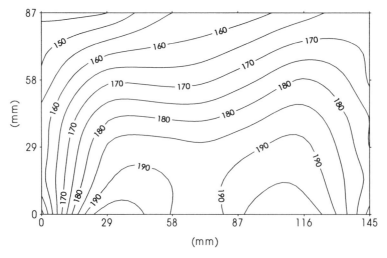

Fig. 4.10. Temperature distribution (no. 8 injection position = 80, runner sectional ratio = 2.4, runner injection angle = 70).

casting operation and theoretical analysis. It can be seen from the tables that the average errors are within 10%. It shows that the models fit the experiments.

5. The Residual Stresses of Die-Casting Die

5.1. *The experimental setup*

In traditional die designing, the runner and body part (cavity block) are not separated. This paper is to design and to make the runner and cavity block separately, in

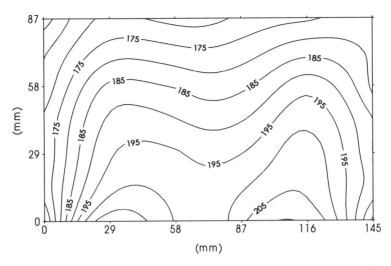

Fig. 4.11. Temperature distribution (no. 9 injection position = 120, runner sectional ratio = 2.4, runner injection angle = 70).

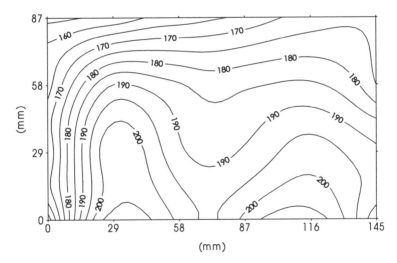

Fig. 4.12. Temperature distribution (no. 10 injection position = 100, runner sectional ratio = 2.4, runner injection angle = 45).

order to know the die casting results under different die casting conditions. That is, the twenty-three runner insert blocks have been put in the moving cavity block respectfully to conduct die casting tests. According to the design of die casting mold, there are twenty-three experiment sets carried out. Figure 5.1 and Table 5.1 show the designs of the die and the runner blocks.

In this study, first, install the cavity block of die casting die into the die block, and then put them on the die casting machine together. After that, set the condition

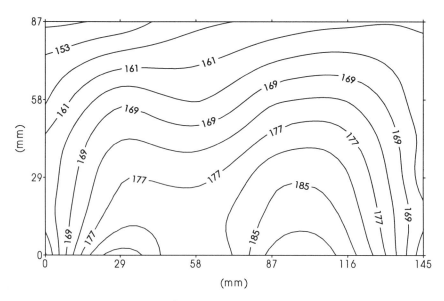

Fig. 4.13. Temperature distribution (no. 11 injection position = 100, runner sectional ratio = 2.4, runner injection angle = 90).

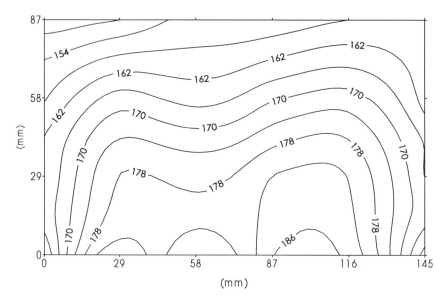

Fig. 4.14. Temperature distribution (no. 12 injection position = 100, runner sectional ratio = 1.8, runner injection angle = 70).

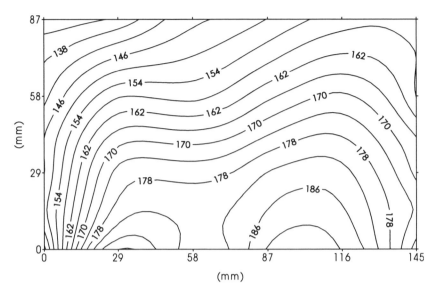

Fig. 4.15. Temperature distribution (no. 13 injection position = 100, runner sectional ratio = 3.0, runner injection angle = 70).

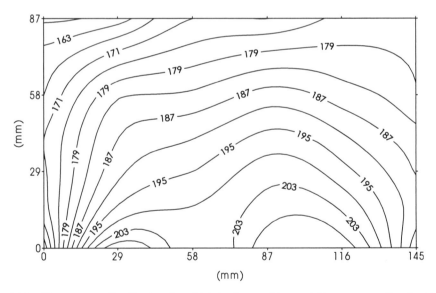

Fig. 4.16. Temperature distribution (no. 14 injection position = 100, runner sectional ratio = 2.4, runner injection angle = 70).

of die casting experiment (e.g. casting pressure: $1200 \, \text{km/cm}^2$, low injection velocity: $0.2 \, \text{m/s}$, high injection velocity $2.8 \, \text{m/s}$, cycling time of casting: 3 shots/min). After the molten aluminum has been injected into the die cavity, it solidifies because of its losing heat. Too high or too low die temperature will have bad effects on the

Table 4.3.   The coefficient of thermal conductivity (cgs).

Temperature (°C)	0	400	660.1	700	400
Data		0.57	0.57	0.22	0.247	25.4

Table 4.4.   The coefficient of thermal expansion ($\times 10^6$).

Temperature (°C)	0	20	200	300	400
Data		23.9	23.9	24.3	25.3	26.49

Table 4.5.   The maximum deformation (mm).

Set no.	1	2	3	4	5
Deformation	0.3464	0.3435	0.3210	0.3377	0.3701
Set no.	6	7	8	9	10
Deformation	0.3880	0.3706	0.3683	0.3958	0.3914
Set no.	11	12	13	14	
Deformation	0.3509	0.3485	0.3463	0.3937	

casting quality. It is very important to know how to control the temperature of die-casting and it's very difficult. First, it has to measure the temperature of die cavity surface, and this research chooses the position under the die cavity for temperature measuring, and put one end of thermal-couple line at the point under the cavity of moving die and fixed die (points *A–H* as show in Fig. 4.1). The other end of the thermal-couple line connects with the GPIB 556 temperature measuring equipment. As the operation of die-casting is undergoing, start the GPIB 556 temperature equipment to measure the temperature of the points under die cavity while the die cavity is opened and closed.

In the process of die casting, the molten metal inject form the runner entry at first, it will be caused high pressure and high temperature. At last, when the molten metal enters into die cavity, the temperature and pressure will be reduced. Due to the function of die casting pressure and high temperature of molten metal, residual stress on the runner is easily caused in the runner part. If the runner is designed improperly, the residual stresses will increase and life of die will shorten. In order to know the changes of residual stresses on the different functions of die casting factors, residual stresses on the runner are measured. There are two positions to be measured: one is the runner entry, and the other is placed near the gate. (The measured places are position marked *RS* shown in Fig. 4.1) Before measuring, mark a sign on the same place of the test runner blocks, stick a strain gauge on the marked position, drill a hole and then measure the residual stresses. The drilling equipment is RS-200 Style. The measuring system is SYSTEM-4000 Style.

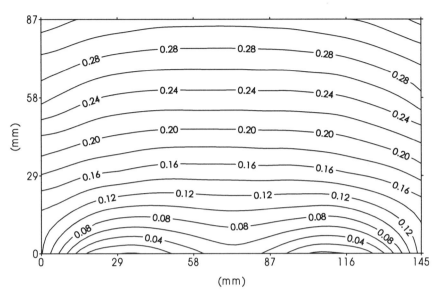

Fig. 4.17.   Deformation (no. 1 injection position = 90, runner sectional ratio = 2.0, runner injection angle = 60).

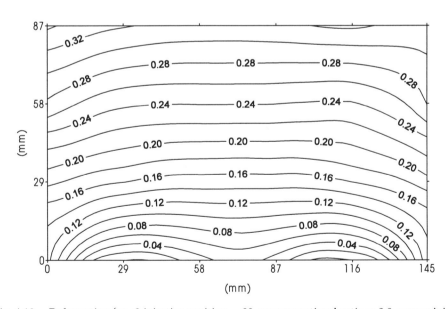

Fig. 4.18.   Deformation (no. 2 injection position = 90, runner sectional ratio = 2.8, runner injection angle = 60).

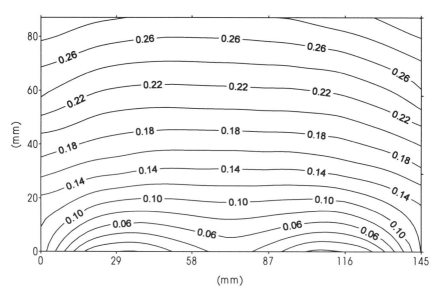

Fig. 4.19. Deformation (no. 3 injection position = 90, runner sectional ratio = 2.0, runner injection angle = 80).

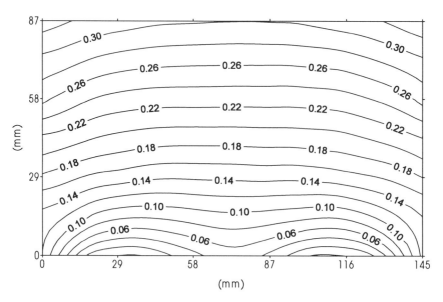

Fig. 4.20. Deformation (no. 4 injection position = 90, runner sectional ratio = 2.4, runner injection angle = 80).

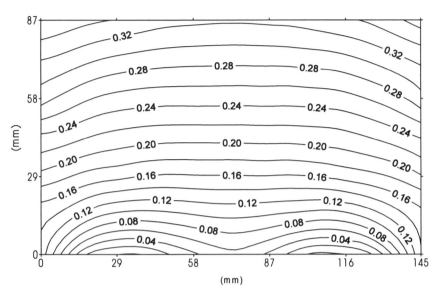

Fig. 4.21. Deformation (no. 5 injection position = 110, runner sectional ratio = 2.0, runner injection angle = 60).

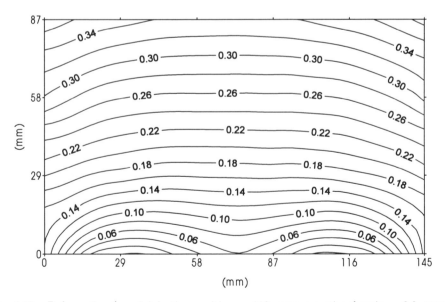

Fig. 4.22. Deformation (no. 6 injection position = 110, runner sectional ratio = 2.8, runner injection angle = 60).

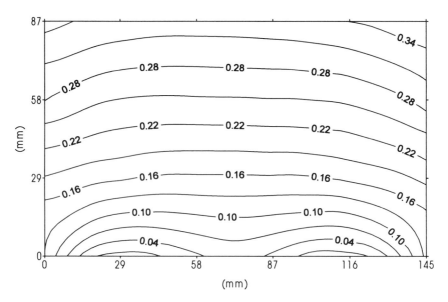

Fig. 4.23. Deformation (no. 7 injection position = 110, runner sectional ratio = 2.8, runner injection angle = 80).

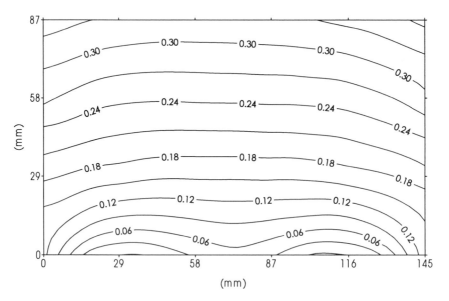

Fig. 4.24. Deformation (no. 8 injection position = 80, runner sectional ratio = 2.4, runner injection angle = 70).

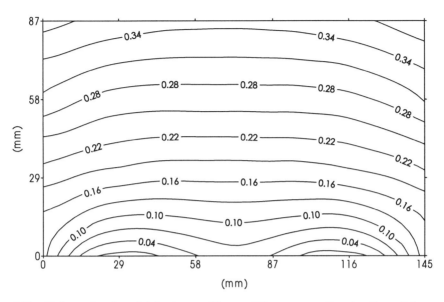

Fig. 4.25. Deformation (no. 9 injection position = 120, runner sectional ratio = 2.4, runner injection angle = 70).

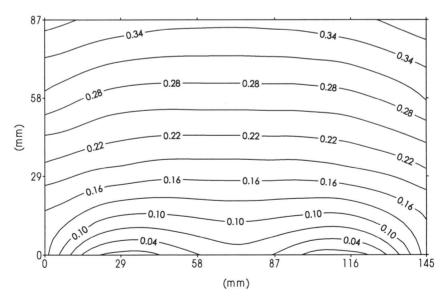

Fig. 4.26. Deformation (no. 10 injection position = 100, runner sectional ratio = 2.4, runner injection angle = 45).

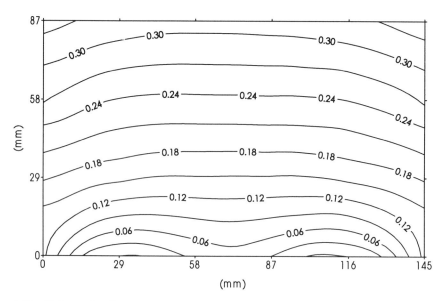

Fig. 4.27. Deformation (no. 11 injection position = 100, runner sectional ratio = 2.4, runner injection angle = 90).

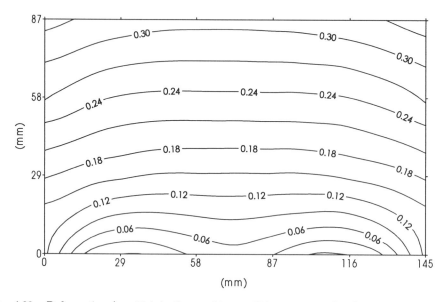

Fig. 4.28. Deformation (no. 12 injection position = 100, runner sectional ratio = 1.8, runner injection angle = 70).

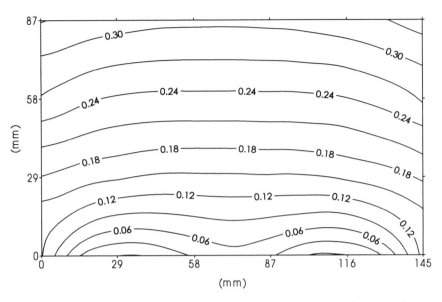

Fig. 4.29. Deformation (no. 13 injection position = 100, runner sectional ratio = 3.0, runner injection angle = 70).

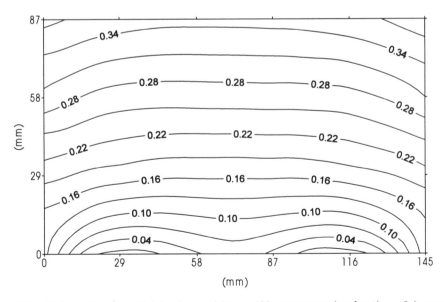

Fig. 4.30. Deformation (no. 14 injection position = 100, runner sectional ratio = 2.4, runner injection angle = 70).

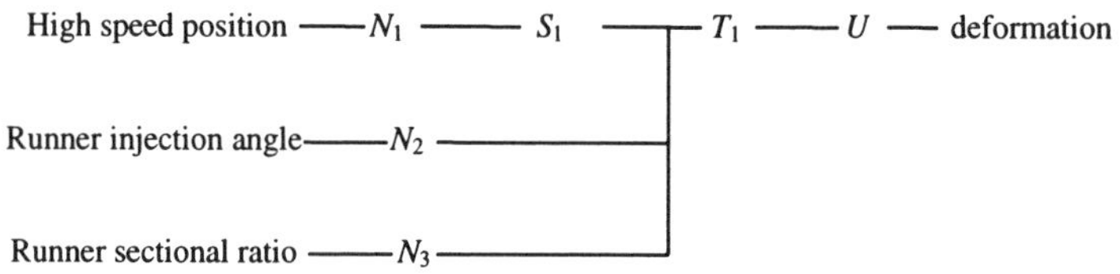

High speed position ——N_1 —— S_1 ——┐ T_1 —— U —— deformation

Runner injection angle——N_2 ——┤

Runner sectional ratio ——N_3 ——┘

$N_1 = -9.27 + 0.0933x_1$

$N_2 = -6.01 + 0.0872x_2$

$N_3 = -6.48 + 2.67x_1$

$S_1 = 1.06x_1 - 0.209x_2$

$T_1 = 1.14 - 1.06x_1^2 - 0.722x_3^2 - 0.355x_1x_2x_3 + 1.61x_1^3$

$U = 0.365 + 0.0245x_1$

Fig. 4.31.   Abductive networks for predicting maximum deformation.

Table 4.6.   The result of a comparison between the experimental maximum deformation and model predicted values.

Name	High speed injected position (mm)	Runner injected angle (°)	Runner sectional ratio	Theory value (mm)	Theory error percent (%)	Experimental measured error (mm)	Predict value (mm)	Predict error percent (%)
Value	110	80	2	0.3652	8	0.397	0.3837	3.4

Although it has methods to measure residual stresses, this experiment has adopted the drilling method that is one of the most popular method used nowadays. The advantages of drilling method include its precision and reliability. This method is often adopted by the American Society for Testing Material (ASTM). According to the principles set by the ASTM, the formula for calculating the stress of blind drilling method is as follows:

$$\sigma_{1,2} = \frac{\varepsilon_1 + \varepsilon_2}{4\bar{A}} \pm \frac{\sqrt{2}}{4\bar{B}}\sqrt{(\varepsilon_1 - \varepsilon_3)^2 + (\varepsilon_2 - \varepsilon_3)^2} \tag{5.1}$$

$$\beta = \tan^{-1}\left(\frac{\varepsilon_3 - 2\varepsilon_2 - \varepsilon_1}{\varepsilon_3 - \varepsilon_2}\right) \tag{5.2}$$

$$\bar{A} = -\left(\frac{1+\nu}{2E}\right)\bar{a} \tag{5.3}$$

$$\bar{B} = -\left(\frac{1}{2E}\right)\bar{b} \tag{5.4}$$

where σ_1 is maximum principle stress, σ_2 is minimum principle stress; ε_i is the measured strain relaxation; ν is Poisson ratio; E is Young modules; a and b are the dimensionally hole-drilling calibration constants; and β is the angle between gauge 1 and principle stress.

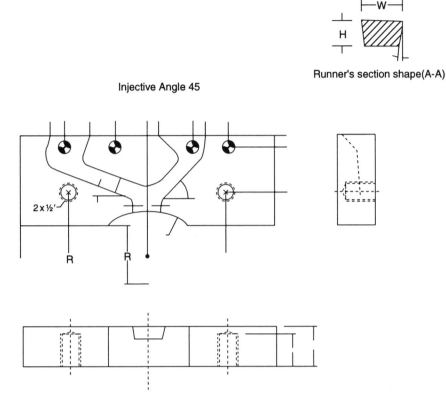

Fig. 5.1.   Insert runner block.

5.2. *Residual stresses*

Generalized development in the modeling of die-casting, needs to have a database and a good relationship to the process parameters and casting results. The precise results and casting process parameters can help to build an exact modeling with the abductive network.

During the die casting process, the molten metal flows from the runner, passes through the gate and enters into die cavity. Therefore, as it is getting nearer to the gate and runner entry, the temperature of the die cavity surface is getting higher. Of course, it also depends on factors such as runner injection angle and design of the gate position.

The residual stresses influence the life of die. Because of the thermal stresses of die casting, die are caused by the heat source of the casting. The highest point of die-casting is getting nearer to the gate and runner entry. This paper, in order to know the residual stresses caused during die casting experiment by different parameters to build the abductive network, the residual stresses only at runner entries and the position near gates of twenty-three runner blocks have been measured.

Table 5.1.　The design of die and runner blocks.

Type	W	H	W/H	θ
1	18	10	1.8	45
2	18	9	2	45
3	18	7.5	2.4	45
4	18	6.5	2.8	45
5	18	6	3	45
6	18	10	1.8	60
7	18	9	2	60
8	18	7.5	2.4	60
9	18	6.5	2.8	60
10	18	6	3	60
11	18	10	1.8	70
12	18	9	2	70
13	18	7.5	2.4	70
14	18	6.5	2.8	70
15	18	6	3	70
16	18	10	1.8	80
17	18	9	2	80
18	18	7.5	2.4	80
19	18	6.5	2.8	80
20	18	6	3	80
21	18	10	1.8	90
22	18	9	2	90
23	18	7.5	2.4	90
24	18	6.5	2.8	990
25	18	6	3	90

Table 4.1 shows the variation of stresses at the runner entries and near the gate with different high-speed plunger injection position, the runner injection angles, and the runner section ratios. The residual stress at places near gate is about 100 Mpa larger than residual stresses at the runner entry. This is why the gate of die is always damaged first. It is worthwhile to mention that the largest residual stress value measured in this experiment always exists in the position within 0.1 mm under the surface layer.

The quality of casting is very important for die casters. Defects of casting are always caused by unsuitable die casting conditions, bad die designing, improper die casting operation, and other human mistakes. Generally, the casting defects are indicated in the porosity, miss-run, shrinkage porosity, cold shut, and soldering. In these experiments, there are more obvious casting defects in shrinkage porosity and cold shut. Table 5.2 shows the situations of casting defects, such as shrinkage porosity and cold shut and understates experimental parameters.

5.3. *Stress model of the experiment*

Based on the developed training database, a three-layer abductive network for predicting the temperature and residual is automatically synthesized.

Table 5.2.   The defect degree of die casting shrinkage porosity and cold shut under different experimental parameters.

No.	High speed injected position (mm)	Runner injected angle (°)	Runner sectional ratio (W/H)	Shrinkage porosity defect	Cold shut defect
1	90	60	2	○	○
2	90	60	2.8	×	○
3	90	80	2	○	×
4	90	80	2.8	◎	○
5	110	60	2	×	○
6	110	60	2.8	◎	◎
7	110	80	2	×	○
8	110	80	2.8	○	○
9	80	70	2.4	○	×
10	120	70	2.4	×	○
11	100	45	2.4	○	◎
12	100	90	2.4	×	○
13	100	70	1.8	○	×
14	100	70	3	○	○
15	100	70	2.4	◎	◎
16	100	70	2.4	◎	◎
17	100	70	2.4	◎	◎
18	100	70	2.4	○	○
19	100	70	2.4	◎	○
20	100	70	2.4	◎	◎
21	100	70	2.4	◎	◎
22	100	70	2.4	○	◎
23	100	70	2.4	◎	○

◎: Least serious; ○: Less serious; ×: Serious.

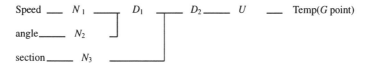

$$N_1 = -9.35 + 0.0935x_1$$
$$N_2 = -6.1 + 0.0876x_1$$
$$N_3 = -6.35 + 2.65x_1$$
$$D_1 = -0.0928 + 0.646x_1 + 0.0994x_1^2 - 0.0735x_1^3$$
$$D_2 = 1.21 + 0.386x_1 - 0.612x_2 + 1.29x_3 - 2.48x_1^2 - 0.0954x_2^2 - 0.0954x_2^2$$
$$\quad - 0.664x_3^2 + 2.23x_1^3 + 0.109 * x_2^3 - 0.723x_3^3$$
$$U = 145 + 9.5x_1$$

Fig. 5.2.   Abductive network for predicting the temperature of G point.

It also accurately predicts the temperature and residual stresses. All polynomial equations used in these networks are listed in Figs. 5.2–5.5. Predict results use abductive network and the results of measure in Tables 5.3 and 5.4.

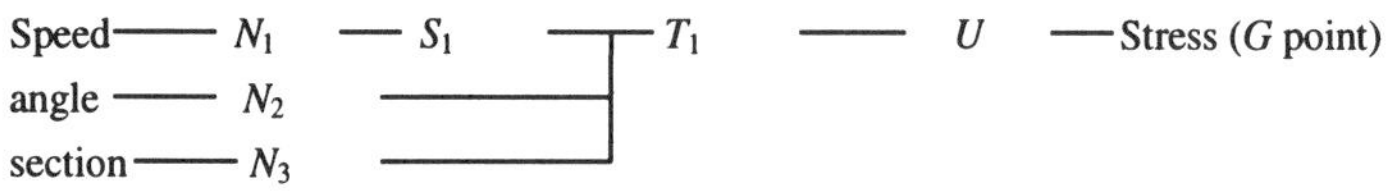

Speed —— N_1 — S_1 ——┐ T_1 —— U —— Stress (G point)
angle —— N_2
section —— N_3

$N_1 = -6.35 + 2.65x_1$
$N_2 = -9.35 + 0.0935x_1$
$N_3 = -6.1 + 0.0876x_1$
$S_1 = -0.289x_1^3$
$T_1 = -1.07 + 1.0x_1 + 0.416x_2 + 1.01x_1^2 + 0.569x_2^2 + 0.349x_3^2 + 0.103x_2^3 + 0.112x_3^3$
$U = 444 + 36x_1$

Fig. 5.3.   Abductive network for predicting the residual stress of G point.

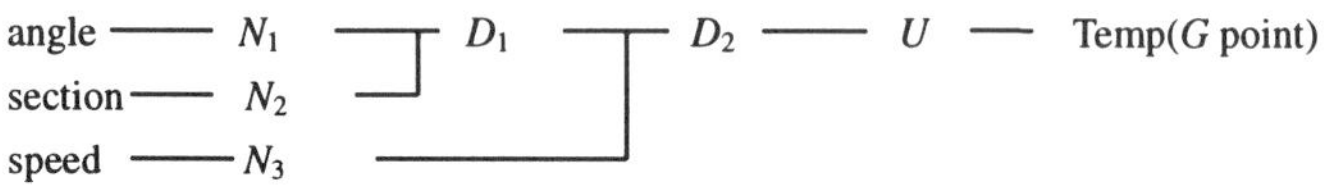

angle —— N_1 ——┐ D_1 ——┐ D_2 —— U —— Temp(G point)
section —— N_2 ——┘
speed —— N_3

$N_1 = -6.1 + 0.0876x_1$
$N_2 = -6.35 + 2.65x_1$
$N_3 = -9.35 + 0.0935x_1$
$D_1 = 0.902x_1^3 - 0.267x_2^3$
$D_2 = -0.591 + 1.07x_1 + 0.85x_1^2 + 0.45x_2^2 + 0.274x_2^3$
$U = 336 + 38.4x_1$

Fig. 5.4.   Abductive network for predicting the residual stress of B point.

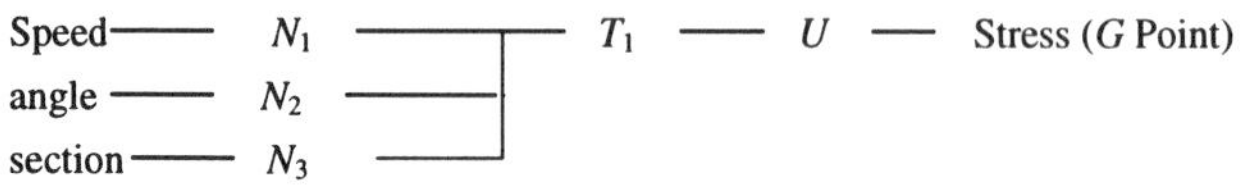

Speed —— N_1 ——┐ T_1 —— U —— Stress (G Point)
angle —— N_2
section —— N_3

$N_1 = -9.35 + 0.0935x_1$
$N_2 = -6.1 + 0.0876x_1$
$N_3 = -6.35 + 2.65x_1$
$T_1 = 1.37 + 0.912x_1 - 0.323x_2 + 0.223x_3 - 0.184x_1^2 - 0.384x_2^2 - 0.928x_3^2$
$\qquad - 0.183x_1^3 + 0.088x_2^3$
$U = 195 + 10.3x_1$

Fig. 5.5.   Abductive network for predicting the temperature of B point.

Table 5.5 shows the result of the comparison between residual stress and model predicting values, which are measured through practical die casting operation. It can be seen from the tables that percentages of error are all within 5 percent. It shows that models fit in the experiment.

Table 5.3.   The results of predicted use abductive network and the results of measured (G point).

No	High speed injected position (mm)	Runner injected angle (°)	Runner sectional ratio (W/H)	Temperature of measured (°C)	Residual stresses of the place gate (Mpa)	Temperature of predicted (°)	Residual stresses of predicted (Mpa)
1	90	60	2	135.5	427.3	140	430
2	90	60	2.8	143.9	396.1	150	410
3	90	80	2	128.1	433.2	130	440
4	90	80	2.8	138.2	414.6	140	410
5	110	60	2	147.1	459.8	150	460
6	110	60	2.8	157.5	439.3	160	440
7	110	80	2	136.3	466.4	140	470
8	110	80	2.8	146.1	442.1	150	450
9	80	70	2.4	143.4	428.2	140	430
10	120	70	2.4	157.2	532.9	160	530
11	100	45	2.4	153.5	425.6	160	420
12	100	90	2.4	148.0	471.5	140	470
13	100	70	1.8	147.2	497.5	140	500
14	100	70	3	131.2	413.1	130	410
15	100	70	2.4	158.2	417.2	150	410

The largest predict error is 4.2%.

Table 5.4.   The results of predicted use abductive network and the results of measured (B point).

No	High speed injected position (mm)	Runner injected angle (°)	Runner sectional ratio (W/H)	Temperature of measured (°C)	Residual stresses of the place gate (Mpa)	Temperature of predicted (°)	Residual stresses of predicted (Mpa)
1	90	60	2	186.9	322.7	190	330
2	90	60	2.8	191.3	304.7	190	310
3	90	80	2	175.2	327.5	180	340
4	90	80	2.8	185.3	306.8	180	310
5	110	60	2	200.1	342.4	200	350
6	110	60	2.8	204.5	320.6	210	320
7	110	80	2	193.6	359.6	190	360
8	110	80	2.8	197.5	334.0	220	330
9	80	70	2.4	198.4	303.7	200	300
10	120	70	2.4	208.8	438.4	210	440
11	100	45	2.4	207.5	307.5	210	300
12	100	90	2.4	187.0	365.7	190	340
13	100	70	1.8	184.8	390.2	180	390
14	100	70	3	190.0	304.9	190	310
15	100	70	2.4	209.8	311.7	210	310

The largest predict error is 3.8%.

Table 5.5.   Comparison of the predict model and observed values of residual stresses.

	High speed injected position (mm)	Runner injected angle (°)	Runner sectional ratio (W/H)	Observed values (Mpa)	Predict values (Mpa)	Predict error percent (%)
The place of runner entry (B point)	110	75	1.9	346.4	370	6.8
The place near gate (G point)	110	75	1.9	452.3	484.3	7.1

6. Choosing of the Injection-Gate and Part-Plane

6.1. *Three dimension flow mold analysis*

Design of die consist of design of runner channel (shape, size and angle), design of die part-plane (selection of die-casting gate), analytical design of life span of the die (residue stress), cooling system etc. The purpose of this chapter is to find the optimal injection gate of die-casting die to cast the thin-shell work-piece. Assumption of casting condition are: die-casting pressure is $1200 \, kg/cm^2$, casting speed is $2.8 \, m/s$, die-casting cycling time is 3 shots/min, pre-heat temperature of the die is $200°C$, injection temperature is $760°C$. And the basic assumption for the flow in the cavity are: (1) 3D-flow; (2) Newton fluid; (3) Laminar flow; (4) Uncompressible fluid; (5) Zero speed of fluid at vertical and horizontal wall direction.

According to the different size and thickness of the work piece used in this study, the symmetrical of the part consist of 17 injection gate set on the surface, and the basic configuration is a $3D$ work piece based on the model flow analysis of $3D$-flow done on COSMOS/M software. The software COSMOS/M (version 2.6) and the pre- and postprocessor GEOSTAR are the Structural Research and Analysis Corporation (SRAC). Utilizing the NONLINEAR and 3D-FLOW model analyzes the change of the heat-stress and heat-strain under the THERMAL-LOADS condition, their mechanical property excepted viscosity, thermal convection coefficient, Yang's coefficient, specific heat, etc. Tables 4.3 and 4.4 shows the modular table of its relation with temperature.

Finite-element used $3D$ and 8-nodes element according to actual $3D$ structure. Divide into three zones on X-direction on the plane, and so for the Y and Z directions. Then introduced with material property of heat transmission coefficient, viscosity coefficient etc., and set the surface temperature as pre-heat temperature ($200°C$) of the die, the injection direction are as stated above, and the temperature of the molten metal is $760°C$.

Select the position of injection gate at the ratio of height, length and width as 0, 1/2 and 1 (work-piece size/injection position), at a total number of 17 points, as illustrated in Fig. 4.1 (Fig. 4.1 show the die-cavity of part). If the injection direction is Z and upward, and located at the lowest point (the ratio of height is zero).

The metal-flow is partially along Y direction, and partially along X direction, and when the XY plane is full, it flow to Z positive direction, until the cavity is completely filled. And the velocity of die-casting is one unit-velocity (one unit-velocity = 2.8 m/s). There die-casting of other position should be $Z = 0$ and fill the X, Y direction, and then fill upward subsequently.

During the process of die-casting study, the overall actual conditions should be considered, including molten liquid zone, solid-molten liquid zone and solid zone three area. Bossing theoretical value include Darcy's flow, Newtonian and Laminar flow three portions, using Mix theory[25,26] governing equations. The governing equations are:

$$\frac{\partial}{\partial x_j}(\rho u_j) = 0 \tag{6.1}$$

Momentum balance,

$$\frac{\partial}{\partial x_j}(\rho u_j u_i) = \frac{\partial}{\partial x_j}\left(\mu \frac{\partial u_i}{\partial x_j}\right) + \frac{\partial}{\partial x_j}\left(\mu \frac{\partial u_j}{\partial u_i}\right) - \frac{\partial P}{\partial x_j}, \quad i = 1, 2, 3; \ j = 1, 2, 3 \tag{6.2}$$

$$\frac{\partial}{\partial x_j}(\rho u_j i) = \frac{\partial}{\partial x_j}\left(\frac{k}{c_p}\frac{\partial u_i}{\partial x_j}\right) - \frac{\partial}{\partial x_j}(\rho u_j \Delta H), \quad j = 1, 2, 3 \tag{6.3}$$

H is the total sensible heat,

$$i = c_p T,$$
$$H = c_p T + \Delta H \tag{6.4}$$
$$\Delta H = \frac{L}{2} + L\frac{i - (1/2)c_p(T_l + T_s)}{c_p(T_l - T_s)} \tag{6.5}$$

T_s and T_l are the temperatures of solid and liquid.

During the analytical process, because the piece is a regular and symmetric object, there are only analytical 17-injection positions selected, and every injection-point contain (1) injection along X-axis direction, (2) injection along Y-axis direction, (3) injection along $+Z$ direction, and (4) injection along $-Z$ direction.

The method for deformation analysis is identical. The set injection temperature are all 760°C, so to maintain a temperature of 760°C at injection gate, and other temperature in using finite elements analysis, the temperature at the instant of filling up is shown in Fig. 6.2. That is, temperature distribution of part at injection liquid filled time, packing time and cooling time can be calculated.

The deformation analysis is to use the solid model analyzed by 3D-flow and nonlinear condition, add the finding temperature to boundary condition required while performing the analysis. Input the configured temperature of each node as initial condition, and make the injection gate as bound constrain condition, to serve as 3D thermal strain conditions and mechanical property change which accompanied the change of temperature. It is an analysis of non-linear stable stage, as shown in Tables 4.3 and 4.4, based on the result of the aforesaid cavity flow analysis perform

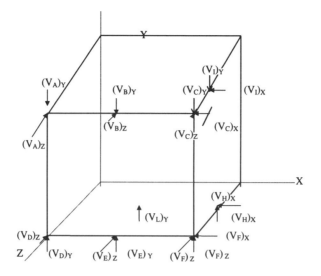

Fig. 6.1.   The position of injection gate.

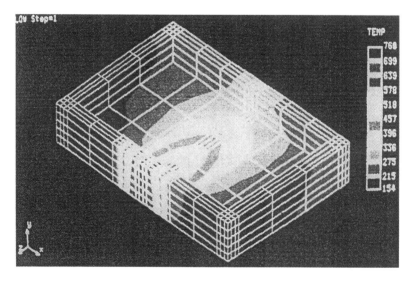

Fig. 6.2.   The temperature distributing of the die-cavity at the instant of filling up.

strain analysis portion. Figure 6.3 shows the deformation profile of part. Table 6.1 shows the deformation result yield from various casting parameter simulations.

6.2. *Creating the deformed model*

The design of die-casting should consider the component system design factors such as runner, injected pressure, injected position, and the material of the die, especially the mold joint of the lower-die and upper-die is very important. Copious experience

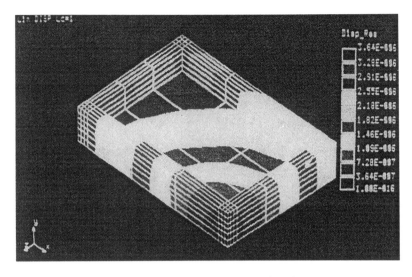

Fig. 6.3. The deformation profile of workpiece.

is essential in manual assessment. If the design under manual assessment is defective, a great deal of time will be wasted in modifying the die, thus minimizing human negligence. Using adductive network to model all design parameters is the safest and most reliable method.

Prior to develop a die-casting deformed model, generally a database had to be trained, and good relationship must exist between the parameters of die-casting process and die-casting die. The correct result and die-casting parameters are helpful for establishing a precise model.

Reasoning from initial facts to new facts with certainty using general principles is called deductive reasoning. However, reasoning in a real situation is very uncertain. Therefore, another class of reasoning using an abductive network is defined as reasoning from initial facts to new facts under uncertainty using general principles.

Miller conducted an important series of experiments concerning human behavior. The research indicates that the human mind processes information by accepting only a finite amount of information at one time. The information is then summarized and then passed on to a higher reasoning level for further processing. In a similar way, an abductive network decomposes input information into smaller, simpler subsystems that can be further sub-divided into several layers using polynomial function nodes. These nodes evaluate a finite number of inputs by a polynomial function and generate an output that serves as the input for subsequent nodes at the next layer.

This study using abductive network predict. Its basic information are shown in Fig. 6.4 — comprising 8 inputs and 1 output and each abductive network node, with PSE = 0.025 obtained from trial and error practice.

Table 6.1.   The maximum deformation of the work-piece.

X (length)	Y (height)	Z (width)	T (thickness)	V_x (vel. of X-direct)	V_y (vel. of Y-direct)	V_z (vel. of Z-direct)	Inject. pos. ratio of X dir.	Inject. pos. ratio of Y dir.	Inject. pos. ratio of Z dir.	Deformation of max. ($\times 0.001$)
30	30	30	3	0	−1	0	1	1	0.5	0.281
30	30	30	3	−1	0	0	1	0	0.5	0.0482
30	30	30	3	0	1	0	1	0	0.5	0.0678
30	30	30	3	−1	0	0	1	0	1	0.0549
30	30	30	3	0	−1	0	1	1	1	0.0782
30	30	30	3	0	0	−1	0.5	1	1	0.0482
30	30	30	3	0	−1	0	0.5	1	1	0.0285
30	30	30	3	0	1	0	0.5	1	1	0.8678
30	30	30	3	0	1	1	1	0	1	0.0846
30	30	30	3	0	0	−1	1	0	1	0.0782
30	30	30	3	0	0	−1	0	0	1	0.0546
30	30	30	3	0	1	0	0	0	1	0.0782
30	30	30	3	−1	−1	0	1	0	1	0.0782
30	30	30	3	0	0	0	1	1	0	0.0546
30	30	30	3	0	−1	0	1	0	0	0.0782
30	30	30	3	0	1	0	1	0	0	0.0846
60	60	60	6	0	−1	0	1	1	0.5	3.96
60	60	60	6	−1	0	0	1	0	0.5	4.09
60	60	60	6	0	1	0	1	0	0.5	4.06
60	60	60	6	−1	0	0	1	0	1	4.84
60	60	60	6	0	−1	0	1	1	1	5.15
60	60	60	6	0	0	−1	0.5	1	1	4.06
60	60	60	6	0	−1	0	0.5	0	1	3.96
60	60	60	6	0	1	0	0.5	1	1	4.06
60	60	60	6	0	1	0	1	0	1	4.81
60	60	60	6	0	0	−1	1	0	1	4.84
60	60	60	6	0	0	−1	0	0	1	4.84
60	60	60	6	0	1	0	0	0	1	4.81
60	60	60	6	0	0	0	0	1	1	4.81
60	60	60	6	0	1	0	0	0	1	5.15
60	60	60	6	−1	−1	0	1	1	0	4.84
60	60	60	6	0	1	0	1	0	0	4.81

5.32	0.5	1	1	0	-1	0	6	100	60	60
5.34	0.5	0	1	0	0	-1	6	100	60	60
5.39	0.5	0	1	0	1	0	6	100	60	60
5.97	1	0	1	0	0	-1	6	100	60	60
6.73	1	1	0.5	0	-1	0	6	100	60	60
5.39	1	0	0.5	-1	0	0	6	100	60	60
5.32	1	1	0.5	0	-1	0	6	100	60	60
5.39	1	0	1	0	1	0	6	100	60	60
6.05	1	0	1	0	1	0	6	100	60	60
5.97	1	0	0	-1	1	0	6	100	60	60
5.97	1	0	0	-1	0	0	6	100	60	60
6.05	1	0	0	0	0	0	6	100	60	60
6.73	0	1	1	0	1	0	6	100	60	60
5.97	0	0	1	0	1	-1	6	100	60	60
6.73	0	1	1	0	-1	0	6	120	120	90
6.05	0.5	0	1	0	0	0	4.5	120	120	90
7.6	0.5	1	1	0	-1	0	4.5	120	120	90
7.89	0.5	0	1	0	0	-1	4.5	120	120	90
7.85	1	0	0.5	0	1	0	4.5	120	120	90
8.33	1	0	0.5	0	0	-1	4.5	120	120	90
9.14	1	1	0.5	0	1	0	4.5	120	120	90
6.65	1	0	1	-1	0	-1	4.5	120	120	90
6.55	1	1	1	0	-1	0	4.5	120	120	90
6.66	1	0	0	0	1	0	4.5	120	120	90
8.35	1	0	0	0	1	0	4.5	120	120	90
8.38	1	0	0	-1	1	0	4.5	120	120	90
8.35	1	0	1	-1	0	0	4.5	120	120	90
9.14	1	0	1	0	0	0	4.5	120	120	90
8.33	0	1	1	0	1	0	4.5	120	120	90
9.14	0	0	1	0	1	-1	4.5	120	120	90
8.35	0	1	1	0	-1	0	4.5	120	120	90
7.89	0.5	0	1	0	0	0	4	120	90	45
8.06	0.5	1	0.5	0	-1	0	4	120	90	45
7.96	0.5	0	0.5	0	1	0	4	120	90	45
7.30	1	1	1	0	1	0	4	120	90	45
8.48	1	0	1	0	1	0	4	120	90	45
8.31	1	0	0	-1	0	0	4	120	90	45

Table 6.1.   Continued

X (length)	Y (height)	Z (width)	T (thickness)	Vx (vel. of X-direct)	Vy (vel. of Y-direct)	Vz (vel. of Z-direct)	Inject. pos. ratio of X dir.	Inject. pos. ratio of Y dir.	Inject. pos. ratio of Z dir.	Deformation of max. (×0.001)
45	90	120	4	0	-1	0	0	1	1	9.76
45	90	120	4	0	-1	0	1	1	0	9.76
90	150	120	5	-1	0	0	1	0	0.5	7.62
90	150	120	5	-1	0	0	0.5	0	1	9.94
90	150	120	5	0	0	-1	0.5	0	1	9.08
90	150	120	5	0	0	0	1	0	1	9.02
90	150	120	5	0	1	0	0	0	1	9.94
90	150	120	5	0	0	-1	1	0	1	8.87
90	150	120	5	-1	1	0	1	0	0	9.94
90	150	120	5	0	1	0	1	0	0	9.87
40	80	80	3.5	0	1	0	1	0	0.5	5.2
40	80	80	3.5	-1	0	0	1	0	1	5.54
40	80	80	3.5	0	-10	1	1	1	6.1	
40	80	80	3.5	0	0	-1	0.5	0	1	4.43
40	80	80	3.5	0	-1	0	0.5	1	1	4.36
40	80	80	3.5	0	1	0	0.5	0	1	4.43
50	100	40	5.5	0	1	0	1	0	1	5.55
50	100	40	5.5	-1	-1	0	1	1	0.5	5.01
50	100	40	5.5	0	1	0	1	0	1	5.59
50	100	40	5.5	0	0	-1	0.5	0	1	3.55
50	100	40	5.5	-1	1	0	0	0	1	5.42
70	110	50	4.5	0	1	0	1	0	0.	5.58
70	110	50	4.5	0	-1	0	1	0	0	5.35
70	110	50	4.5	0	0	0	1	0	0.5	5.47
70	110	50	4.5	-1	-1	0	1	1	1	6.57
70	110	50	4.5	0	0	0	0.5	1	1	4.45
75	40	75	45	0	0	0	1	0	1	6.47
75	40	75	5.0	0	-1	0	1	1	1	6.68
75	40	75	5.0	0	1	-1	0	0	1	6.4
75	40	75	5.0	-1	0	0	1	0	0.	4.12
75	40	75	5.0	-1	0	0	1	0	0.5	5.2
75	40	75	5.0	0	0	-1	0.5	0	1	4.84
75	40	75	5.0	0	1	0	0.5	0	1	4.81
75	40	75	5.0	0	0	-1	1	0	1	5.21

5.19	1	0	0	0	1	0	5.0	75	40	75
5.20	0	0	1	0	0	-1	5.0	75	40	75
5.71	0	1	1	0	-1	0	5.0	75	40	75
9.93	0.5	1	1	0	-1	0	6.0	150	150	150
10.03	0.5	0	1	0	0	-1	6.0	150	150	150
10.02	0.5	0	1	0	1	0	6.0	150	150	150
12.1	1	0	1	0	0	-1	6.0	150	150	150
12.8	1	1	1	0	1	0	6.0	150	150	150
10.03	1	0	0.5	-1	0	0	6.0	150	150	150
9.93	1	1	0.5	0	-1	0	6.0	150	150	150
10.02	1	0	0.5	0	1	0	6.0	150	150	150
12.1	1	0	1	0	1	0	6.0	150	150	150
12.1	1	0	0	-1	0	0	6.0	150	150	150
12.1	1	0	0	-1	0	0	6.0	150	150	150
12.1	0	0	0	0	1	0	6.0	150	150	150
12.8	0	1	1	0	-1	-1	6.0	150	150	150
12.1	0	0	1	0	0	0	6.0	150	150	150
12.8	0.5	1	1	0	-1	0	6.0	150	150	150
12.1	0.5	0	1	0	1	-1	6.0	150	150	150
5.94	0.5	1	1	0	-1	0	4.5	90	90	90
6.13	1	0	1	0	1	-1	4.5	90	90	90
6.09	1	1	0.5	-1	0	0	4.5	90	90	90
7.27	1	0	0.5	0	1	0	4.5	90	90	90
7.72	1	0	0.5	0	0	0	4.5	90	90	90
6.13	1	0	1	0	-1	0	4.5	90	90	90
5.94	1	1	1	-1	0	0	4.5	90	90	90
6.09	1	0	0	-1	-1	0	4.5	90	90	90
7.22	1	1	0	0	1	0	4.5	90	90	90
7.27	1	0	0	0	1	0	4.5	90	90	90
7.27	1	1	1	0	0	-1	4.5	90	90	90
7.22	1	0	1	0	0	0	4.5	90	90	90
7.72	1	0	1	0	1	0	4.5	90	90	90
7.27	1	0	1	0	-1	0	4.5	90	90	90
7.72	0	0	0	0	0	-1	4.5	90	90	90
7.72	0	1	1	0	-1	0	4.5	90	90	90
7.22	0	0	1	0	1	0	4.5	90	90	90

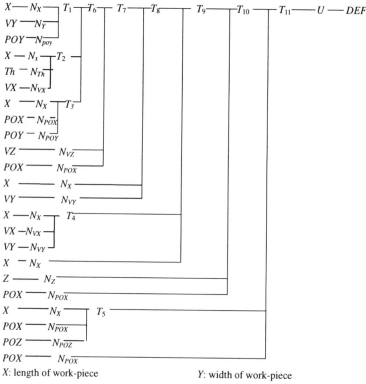

X: length of work-piece   Y: width of work-piece

Z: height of work-piece   Th: thickness of work-piece

POX: injection position of X direction   POY: injection position of Y direction

POZ: injection position of Z direction   VX: injection velocity of X direction

VY: injection velocity of Y direction   VZ: injection velocity of Z direction

DEF: The work-piece max. deformation

$$N_X = -1.13 + 0.0931X_1$$
$$N_Z = -2.2 + 0.0295X_1$$
$$N_{VX} = 2.2 + 0.504X_1$$
$$N_{VY} = 0.359 + 3.08X_1$$
$$N_{VZ} = 0.0194 + 1.35X_1$$
$$N_{POX} = 0.119 + 1.95X_1$$
$$N_{POY} = -1.57 + 2.236X_1$$
$$N_{POZ} = -0.744 + 2.13X_1$$

Fig. 6.4.   Abductive networks for predicting maximum deformation.

In order to indicate whether it is accurate for the established model, a set of die-casting piece which is different from the above stated is served as a model test. The casting parameter is $40 \times 60 \times 80 \times 4\,\text{mm}$, and the die-casting position are at ratio 1, 1, 0 of length, width and height, as shown in Table 6.4. The error between

$N_{Th} = -1.9 + 0.218X_1$

$T_1 = 0.0808 - 0.855X_1 + 0.599X_2 + 0.304X_3 - 0.317X_1^2 + 0.546X_2^2 - 1.04X_3^2 - 0.506X_1X_2$
$\quad - 0.23X_1X_3 + 0.128X_2X_3 + 0.369X_1X_2X_3 + 0.175X_1^3 - 0.82X_3^3$

$T_2 = 0.126 - 2.5X_1 - 0.155X_2 - 0.411X_3 - 2.37X_1^2 + 0.226X_2^2 + 0.888X_3^2 - 0.302X_1X_3$
$\quad + 0.281X_2X_3 + 0.763X_1X_2X_3 + 0.523X_1^3 + 0.248X_3^3$

$T_3 = 0.0843 - 1.04X_1 + 2.23X_2 - 0.117X_3 - 0.377X_1^2 + 0.67X_2^2 - 0.658X_3^2 - 0.64X_1X_2$
$\quad + 0.109X_1X_3 - 0.17X_2X_3 + 0.277X_1X_2X_3 + 0.225X_1^3 + 0.466X_2^2 - 0.438X_3^3$

$T_4 = 0.0425 - 0.207X_1 - 0.758X_2 + 0.514X_3 - 0.234X_1^2 + 0.431X_2^2 + 0.157X_3^2 - 0.593X_1X_3$
$\quad - 0.329X_2X_3 + 0.307X_1X_2X_3 + 0.531X_1^3 + 0.129X_2^3$

$T_5 = -0.243 - 1.21X_1 + 1.8X_2 + 0.527X_3 - 0.798X_1^2 + 0.97X_2^2 - 0.118X_3^2 - 0.1X_1X_2$
$\quad + 0.308X_1X_3 - 1.75X_2X_3 + 0.78X_1X_2X_3 + 0.269X_1^3 - 0.6X_2^3$

$T_6 = -0.0849 - 0.453X_1 + 0.756X_2 + 1.07X_3 - 0.627X_1^2 + 0.564X_2^2 - 0.378X_3^2 - 0.248X_1X_2$
$\quad + 1.31X_1X_3 - 1.39X_2X_3 - 1.29X_1X_2X_3 + 0.546X_1^3 + 0.278X_2^3 + 0.0182X_3^3$

$T_7 = -0.0743 + 1.15X_1 + 0.0683X_2 + 0.737X_3 + 0.11X_1^2 - 0.039X_2^2 + 0.0393X_3^2 - 0.403X_1X_2$
$\quad - 0.0949X_1X_3 + 3.12X_2X_3 + 3.22X_1X_2X_3 - 0.0214X_1^3 - 0.232X_2^3 - 0.203X_3^3$

$T_8 = 0.15 + 0.798X_1 - 0.151X_2 + 0.0492X_3 - 0.22X_1^2 - 0.0592X_1X_3 - 0.303X_2X_3$
$\quad + 0.0289X_1X_2X_3 + 0.611X_1^3 + 0.0417X_2^3$

$T_9 = -0.142 + 1.23X_1 - 0.0783X_2 - 0.0687X_3 + 0.301X_1^2 + 0.164X_2^2 + 0.663X_3^2 - 0.313X_1X_3$
$\quad - 0.382X_1X_3 + 0.718X_2X_3 + 0.0402X_1X_2X_3 - 0.0478X_1^3 - 0.192X_2^3 - 0.0123X_3^3$

$T_{10} = 0.968X_1 + 0.0585X_2 - 0.115X_3 - 0.0124X_1^2 + 0.0349X_2^2 + 0.0602X_3^2 + 0.0101X_1X_2$
$\quad + 0.32X_1X_3 - 0.432X_2X_3 + 0.118X_1X_2X_3 - 0.15X_1^3 - 0.15X_1^3 - 0.227X_2^3 + 0.163X_3^3$

$T_{11} = -0.205 + 1.12X_1 - 0.656X_2 - 0.009X_3 - 0.21X_1^2 + 0.299X_2^2 + 0.0077X_3^2 - 0.0819X_1X_2$
$\quad + 0.0361X_1X_3 + 0.179X_2X_3 + 0.13X_1X_2X_3 - 0.218X_1^3 + 0.15X_2^3 - 0.008X_3^3$

$U = 0.708 + 0.582X_1$

Fig. 6.4. (*Continued*)

the predicted and simulated value is approximately 10%, indicated that this model is suitable for this simulation program.

7. Optimal Die Design for Manufacturing System

7.1. *Die-casting and product part optimization design*

Once the model of the relationship of functions of die conditions and product parts are obtained, this model can be used to find the optimal parameters in the die-casting process. The optimal parameters of process can be obtained by using the objective function to serve as a starting point. The objective function obj is formulated as follows:

$$\text{obj} = [w_{11} \text{ deformation (1 injection gate)}] + [w_{21} \text{ deformation (runner-shape}$$
$$\text{and part-deformation)}] + [w_{31} \text{ temperature (runner shape)} + w_{32} \text{ residual}$$
$$\text{stresses (runner shape)}] \tag{7.1}$$

Table 6.2. The result of a comparison between the simulation maximum deformation and model predicted values.

Name	Length (mm)	Width (mm)	Height (mm)	Thick-ness (mm)	Injection position ratio x direc-tion (length/ posi-tion)	Injection position ratio y direc-tion (length/ posi-tion)	Injection position ratio z direc-tion (length/ posi-tion)	Simula-tion max. defor-mation value (mm × 0.001)	Perdict max. defor-mation value (mm × 0.001)	Perdict error (%) per-cent (%)
Value	50	70	90	4	1.0	1.0	0	5.92	6.11	3.2%

where w_{11}, w_{21} and w_{3n} $[n = 1 \sim 2]$ are weights of normalized injection-positions, and normalized runner-shape and part-deformation and normalized mold-life-parameters in optimization.

7.2. *The optimization position choosing of injection gate*

7.2.1. *Weight function choosing of the injection gate position*

If $w_{1n} \neq 0$, $w_{2n} = w_{3n} = 0$ the simulated annealing algorithm is conducted to search for the optimal injection gate chosen for the die-casting die. Once the maximum deformation of die casting process model has been developed, the use of the model to optimize die-casting processes and to obtain optimal process parameters will be used for setting the objective function starting. As mentioned before, the main effect of the temperature is to cause deformation. Therefore, the permitted deformation decides the design of die. The optimal parameters of process can be obtained by using the objective function to serve as a starting point. The objective function obj is formulated as follows:

$$\text{obj} = w_{11} \text{ deformation (injection position)} \tag{7.2}$$

In the meantime, the deformed, design of the size of die-casting die should meet the simulation data method. That means the basic condition of optimization should fall in a certain range:

(1) The deformation obtained from optimization should be bigger than the minimum deformation, and is smaller than the maximum deformation.
(2) The cavity-die length obtained from optimization should be bigger than the minimum length, and is smaller than the maximum length.
(3) The cavity-die width obtained from optimization should be bigger than the minimum cavity-die width, and is smaller than the maximum cavity-die width.
(4) The cavity-die height obtained from optimization should be bigger than the minimum cavity-die height, and is smaller than the maximum cavity-die height.
(5) The casting thickness obtained from optimization should be bigger than the minimum casting thickness, and is smaller than the maximum casting thickness.

The inequality is given as follows:

the lowest deformation < deformation < the highest deformation   (7.3)

the smallest cavity-die length < length < the largest cavity-die length   (7.4)

the smallest cavity-die width < width < the largest cavity-die width   (7.5)

the smallest cavity-die height < height < the largest cavity-die height   (7.6)

the smallest casting material thickness < casting material thickness

    < the largest casting material thickness   (7.7)

The upper bound conditions should be kept at an acceptable level in finding the optimization runner design.

7.2.2. *Weight function choosing and results discussion*

If $w_{1n} \neq 0$, $w_{2n} = w_{3n} = 0$, then it can find the optimal parameters in the chosen injection position. Once the model of the relationship of functions of injection position is obtained, this model can be used to find the optimal parameters in the milling process. Examples of the simulation are used to illustrate the process of optimizing the die-casting parameters. When the weight function $w_{11} = 1$ (deformation error weighted value = 1), fixed the part with size of $16 \times 64 \times 96 \times 6$ mm. The parameters used in the simulation annealing algorithm are given as follows: the initial temperature $T_s = 100°\text{C}$, the final temperature $T_e = 0.0001°\text{C}$, the decaying ratio $CT = 0.95$, the Boltzmann constant $k_s = 0.00667$, and the upper bound of injected position of die-casting is $30 \times 64 \times 96$ mm, upper bound of material thickness is 6.0 mm, and the lower bound of injected position of die-casting is $0 \times 0 \times 0$ mm, lower bound of material thickness is 3.0 mm. Simulated annealing is used for finding the optimal casting parameter as shown is Table 7.2. These assure the minimum dimensional error from the simulated annealing equation, finds $X = 0$, $Y = 30$, $Z = 0$ as optimal position, and serves as separate line and injection gate.

For the product obtained from designing to formation is shown in Table 7.1. It uses finite element method for deformation analysis on approximate size, and via the learning result of abductive network model. And it is compared with the deformation of the parts, which are physically produced from optimized die-casting piece, the error is within the range of 15%.

Table 7.1. The result of a comparison between the experimental maximum deformation and the optimization model predicted values.

Name	Length (mm)	Width (mm)	Height (mm)	Thickness (mm)	Injection position ratio *x* direction (length/ position)	Injection position ratio *y* direction (length/ position)	Injection position ratio *z* direction (length/ position)	Experimental max. deformation value (mm × 0.001)	Perdict max. deformation value (mm × 0.001)	Optimization predict error percent (%)
Value	30	64	96	6	0	0.5	0	6.25	6.63	6.08%

Table 7.2.   The result of a comparison between the experimental minimum deformation
and the optimization model predicted values.

Name	High speed injected postion (mm)	Runner injected angle (°)	Runner sectional ratio	Experimental measured error (mm)	Predict value (mm)	Predict error percent (%)
Value	87	55	3	0.352	0.321	8.8%

7.3. *The optimal parameters design of die-casting process*

7.3.1. *Weight function choosing of process parameters*

If $w_{1n} = 0$, $w_{2n} \neq 0$, $w_{3n} = 0$, once the maximum deformation of die casting
process model has been developed, the use of the model to optimize die casting
processes and obtain optimal process parameters will be obtained for setting the
objective function starting. As mention before, the main factor of the temperature
is caused by the deformation. Therefore, the deformation decide the design of die.
The objective function obj is formulated as follows:

$$obj = w_{21} \text{ deformation (the relationship of runner shape and part deformation)} \tag{7.8}$$

where w_{21} is weights of normalized runner injected angle, normalized high speed
injected position, normalized runner section ratio and normalized residual stresses
in optimization.

In the meantime, the deformation has to accord the experiment result it need
to higher the lowest-deformation and lower the highest-deformation. The runner
injected angle has to higher the smallest-angle and lower the largest-angle, and the
high speed injected position also has to higher the lowest-position and lower the
highest-position. The inequality is given as follow:

$$\text{the lowest deformation} < \text{deformation} < \text{the highest deformation} \tag{7.9}$$

$$\text{the smallest runner injected angle} < \text{runner injected angle}$$
$$< \text{the largest runner injected angle} \tag{7.10}$$

$$\text{the lowest high speed injected position} < \text{high speed injected position}$$
$$< \text{the highest high speed injected position} \tag{7.11}$$

$$\text{the lowest runner section ratio} < \text{runner section ratio}$$
$$< \text{the highest runner section ratio} \tag{7.12}$$

The upper bound conditions should be kept at an acceptable level in finding the
optimization design of runner.

7.3.2. *The weight function of die-casting processing choosing and results discussion*

Several cases are presented to illustrate the optimization of process parameters in die casting operations. They are only utilized (runner injection position = 95 and runner injection angle = 55) for analysis. In these cases, the parameters used in the simulate annealing algorithm are given as flows: the initial temperature $T_s = 100°$C, the final temperature $T_e = 0.0001°$C, the decaying ratio $C_T = 0.95$, the Boltzmann constant $k_B = 0.00667$, the upper bound of the injection angle is 55 and the runner injection position is 95. Through the simulated annealing searching, the optimal parameters are shown in Figs. 7.1 and 7.2 to keep the minimum deformation. In Fig. 7.1, to fixed high speed injected position is 95 and runner section is 3.0 (the shape of runner, $W = 18$ mm, $H = 6$ mm), to adjust the parameter of runner injected angle. In Fig. 7.2, to fixed the parameter of runner injection is 95 and the runner injection angle is 90, and to adjust the parameter of the runner sectional ratio. In two cases, we get the alikes when the runner injection angle is 90, and the runner sectional ratio is 3.0, which is the optimization condition. In Figs. 7.3 and 7.4, we get the alikes when the injection angle is 55. It is fixed, and to adjust the parameter of the runner sectional ratio and the runner injection position, the runner section is 3.0 (the shape of $W = 18$ mm, $H = 6$ mm) and the injection

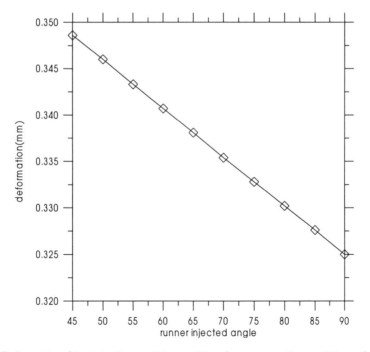

Fig. 7.1.   Deformation (the injection position = 95 and runner section = 3.0 are fixed, when injection angle = 90 has the minimum deformation = 0.3250 mm).

J.-C. Lin

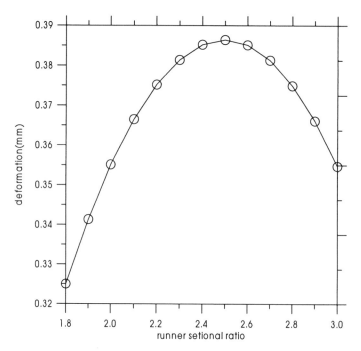

Fig. 7.2. Deformation (the injection position = 95 and the runner injection angle = 90 are fixed, when runner sectional ratio = 3.0 has the minimum deformation = 0.3250 mm).

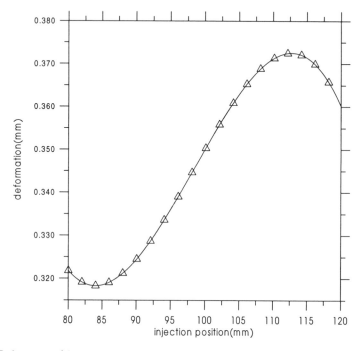

Fig. 7.3. Deformation (the runner section = 3.0 and injection angle = 55 are fixed, when injection position = 87 has the minimum deformation = 0.321 mm).

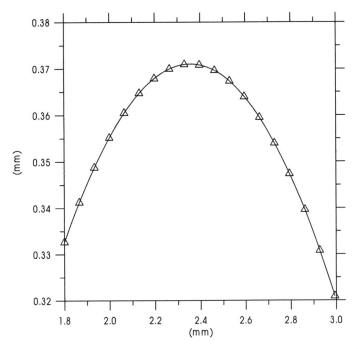

Fig. 7.4. Deformation (the injection angle = 55 and injection position = 87 are fixed, when sectional ratio = 3.0 has the minimum deformation = 0.321 mm).

position is 87, which is the optimum condition. In foregoing discussion, it has been clearly shown that the process parameters with an optimum die casting performance can be systematically obtained through this approach.

Table 7.2 shows the optimization process, when the parameter of injection position of 87 mm, the sectional ratio of 3.0, and injection angle of 55°C — a comparison between the experimental minimum deformation and model predicted values. The errors are within 10 percent. In foregoing discussion, it has been clearly shown that the process parameters for optimum die casting performance can be systematically obtained through this approach.

7.4. The optimize design of die-casting die life

7.4.1. The weight function choosing of die-casting die life parameters

If $w_{1n} = w_{2n} = 0$, $w_{3n} \neq 0$, when the die casting process model has been developed, the use of the model to optimize die casting processes and obtain optimal process parameters will be achieved in setting the objective function starting. As mentioned before, the main factor of the temperature is caused by the residual stress. Therefore, the temperature and residual stress decide the life of die. Since the temperature and residual stress are two different objects, a normalized temperature and stress between zero and one is required. A weighting function is then adopted to transform the normalized temperature and residual stress into a single objective format. The objective function obj is formulated as follows:

$$\text{obj} = w_{31} \text{ Temperature} + w_{32} \text{ stress (runner shape)} \tag{7.13}$$

In the meantime, the temperature has to accord the experiment result it needs to higher the lowest-temperature and lower the highest-temperature. The runner-injected angle has to raise the smallest-angle and lower the largest-angle, and the high-speed injected position also has to raise the lowest-position and to lower the highest-position. The inequality is given as follows:

$$\text{the lowest temperature} < \text{Temperature} < \text{the highest temperature} \tag{7.14}$$

$$\text{the smallest runner injected angle} < \text{runner injected angle}$$
$$< \text{the largest runner injected angle} \tag{7.15}$$

$$\text{the lowest speed injected position} < \text{injected position}$$
$$< \text{the highest speed injected position} \tag{7.16}$$

The upper bound conditions should be kept at an acceptable level when trying to find the optimization design of runner.

7.4.2. The weight function of die-casting die life and results discussion

Several cases are presented to illustrate the optimization of process parameters in die casting operations. There is only a utilized G point to do the analysis because it has the maximum residual stresses. In these cases, the parameters used in the simulate annealing algorithm are given as flows: the initial temperature $T_s = 100°C$, the final temperature $T_e = 0.0001°C$, the decaying ratio $C_T = 0.95$, the Boltzmann constant $k_B = 0.00667$, the upper bound of the high speed injected position is $120\,\text{mm}$, and the runner injected is $90°$. In addition, the sectional ratio selects 2.3, 2.5, and 2.7, different from those used in the training database — using these cases to verify how well the abductive network perform. Through the simulated annealing searching, the optimal parameters are listed in Table 7.3. The process with a low temperature and low residual stresses can keep the die alive. It needs low residual stresses more than low temperature so it can increase the weight of w_{32}. For example, the runner sectional ratio is 2.3, $w_{31} = 0$, and $w_{32} = 1$ in Table 7.3, the high speed injected position is $95.51\,\text{mm}$ and the runner injected angle is $69.1°$. It obtains the lowest residual stresses of $403.1\,\text{Mpa}$ in the optimization procedure. In Table 7.3, as the weight of w_{31} increases the process parameters decrease the temperature but the residual stresses increase. It is not expected to happen. The values of w_{31} can adjust the trade-off between temperature and residual stresses in optimization. The best condition is the selected $w_{31} = 0$.

Figures 7.5 and 7.6 show that when the parameter of runner sectional ratio is fixed and the weight function $w_{31} = 0$, we have two cases. First, fix the high-speed injected position to $95.51\,\text{mm}$ and to adjust the parameter of runner injected angle.

Table 7.3.   The optimization of die casting process and the relationship of w_1, w_2.

No.	High speed injected position (mm)	Runner injected angle (°)	Runner sectional ratio (W/H)	w_1	w_2	Predicted runner temperature (°C)	Predicted runner residual stress (MPa)
1	95.51	69.7	2.3	0	1	144.7	403.1
2	94.68	70.35	2.3	0.5	1	144.1	403.2
3	92	72.5	2.3	2	1	142.4	405.6
4	92.75	90	2.3	1	0	137.9	466
5	95.63	69.67	2.5	0	1	151	402.7
6	94.83	70.36	2.5	0.5	1	150.4	402.8
7	94.24	72.23	2.5	2	1	149.7	403.6
8	92.43	90.0	2.5	1	0	144	466.7
9	96.21	69.68	2.7	0	1	157.3	405.2
10	94.99	70.48	2.7	0.5	1	156.4	405.4
11	93.25	72.5	2.7	2	1	154.4	406.9
12	94.21	90.0	2.7	1	0	148.1	467.3

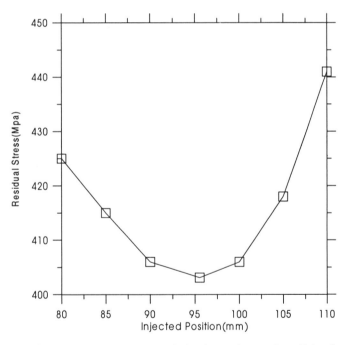

Fig. 7.5.   Effect of die-casting parameters in optimization and general condition ($w_{41} = 0$, $w_{42} = 1$, runner sectional ratio = 2.3, runner adjust injected angle = 69.7°, adjust injected position, the minimum residual stresses is 403.1 Mpa, the injected position 95.5 mm).

Second, fix the parameter of runner injected angle to 69.7° and adjusts the parameter of high speed injected position. We get similar result in $w_{31} = 0$, high speed injected position is 95.51, and the injected angle is 69.7°, which is the optimization condition.

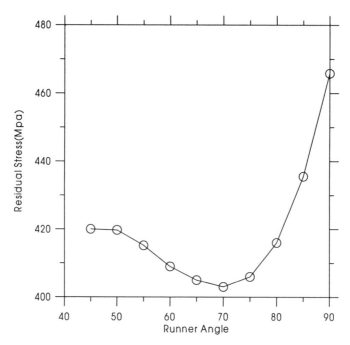

Fig. 7.6.   Effect of die-casting parameters in optimization and general condition ($w_{41} = 0$, $w_{42} = 1$, runner sectional ratio = 2.3, high speed injected position = 95.51 mm, adjust injected angle, the minimum residual stresses 403.1 Mpa, about injected angle 69.1).

In the other case, when the speed injection position is 105 mm, $w_{31} = 0$, and $w_{32} = 1$, when the injection angle is 72.16° and the runner section ratio is 2.85, it obtains the lowest residual stresses of 409 Mpa in the optimization procedure. In other words, when fixing the injection angle, we can get the runner optimization shape — $W = 18$ mm, $H = 6.3$ mm and the runner injection angle $\theta = 72.16°$. Figure 7.7 shows the relationship of deformation and runner section ratio. When fixing the injection angle $\theta = 72.16°$ and the speed injection position 105 mm, you can choose the error in a reasonable range if the die design need a precise dimension.

In previous discussion, it has been clearly shown that the process parameters with an optimum die casting performance can be obtained through this approach.

8. Conclusion

According to the results of this study, the main points can be concluded as follows:

(1) Using runner blocks of insertion-type for a die casting tests can reduce testing time and costs. Because the runner blocks can have different designs if the dies casting test results are poor, one needs only another runner block, which can be obtained easily.

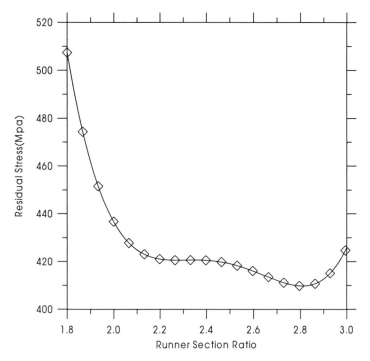

Fig. 7.7. Effect of die-casting parameters in optimization and general condition (runner injection angle = 72.16°, speed injected position is 105 mm, adjust runner section ratio, the minimum residual stresses is 409 Mpa, about runner section ratio 2.85).

(2) Self-organized abductive networks were used to model the die casting process. The relationships between the process parameters of the die casting performance can be constructed by the abductive networks, and the system can predict the die casting performance very accurately.

(3) A gobble optimization algorithm, simulated annealing, is then applied to these networks to obtain optimal die casting process parameters. As a result, a useful technique for modeling making the insertion-style runner blocks has been demonstrated in this study.

References

1. L. W. Garber, Some preliminary comments: Filling of the cold chamber during cold chamber during slow-shot travel, *Die Casting Engineer* **25**, 4 (1981) 36–38.
2. L. W. Garber, Theoretical analysis and experimental observation of air entrapment during cold chamber filling, *Die Casting Engineer* **26**, 3 (1982) 14–22.
3. L. W. Garber, Entrapped air in aluminum die casting, *Die Casting Engineer* **26**, 3 (1982) 14–22.
4. T. P. Groeneveld and W. D. Kaiser, Effect of metal velocity and die temperature on metal-flow distance and casting quality, *Die Casting Engineer* **23**, 5 (1979) 44–49.
5. R. L. Truelove, Die casting temperature control: A new science, *Die Casting Engineer* **21**, 5 (1982) 28–31.

6. T. P. Groeneveld and W. D. Kaiser, Guidelines for locating waterlines, *Die Casting Engineer* **21**, 5 (1977) 14–21.

7. A. B. Draper and J. K. Sprinkle, Waterline location within a die casting die, *Die Casting Engineer* **29**, 4 (1985) 14–21.

8. S. H. Jong, H. Y. Chou, C. R. Li and W. S. Hwang, Application of mold filling analysis in design of die casting die, *Chukung Quarterly* **73** (1992) 1–9.

9. M. Gotoh and Y. Shibata, Elastic-plastic FEM analysis of the heading process and die forging process of gear blank, *Journal of Materials Processing Technology* **27** (1991) 9–110.

10. M. Glowacki and M. Pietrzyk, Experimental substraniation of rigid plastic finite element molding of three-dimensional froming process, *Journal of Materials Processing Technology* **19** (1989) 295–303.

11. K. Mori, K. Osakada and M. Shiom, Finite element modeling of forming processing solid liquid phase, *Journal of Materials Processing Technology* **27** (1991) 111–118.

12. G. S. A. Shawki and A. Y. Kandel, A review of design parameters of machine performance for improve die casting quality, *Journal of Materials Processing Technology* **16** (1988) 315–333.

13. T. Altan, Design and manufacture of die and mold, *Annals of CIRP* **36**, 2 (1987) 455–465.

14. J. Hurt, A taxonomy of CAD/CAE system, *Manufacturing Review* **2** (1989) 170–175.

15. J. Corbett, A CAD-integrate 'Knowledge-based system' for design of die case component, *Annals of CIRP* **40**, 1 (1991) 103–110.

16. G. J. Montgomery and K. C. Drake, Abductive reasoning network, *Neurocomputing* **2** (1991) 97–104.

17. S. Kirkpartick, C. D. Gelatt and M. P. Vecchi, Optimization by simulated annealing, *Science* **220**, 4958 (1983) 671–680.

18. S. Geman and D. Geman, Stochastic relaxation, Gibbs distributions and the Bayesian restoration of images, *IEEE Trans. on Pattern Analysis and Machine Intelligence* **6**, 6 (1984) 721–741.

19. B. W. Lee and B. J. Sheu, *Hardware Annealing in Analog VLSI Neuron Computing* (Kluwer Academic Publishers, London, 1991).

20. C. Zhang and H. P. Wang, The discrete tolerance optimal problem, *ASME Manufacturing Review* **6**, 1 (1991) 60–71.

21. Y. S. Trang, S. C. Ma and L. K. Chung, Determination of optimal cutting parameters in wire electrical discharge machining, *Int. J. Mach. Tools Manufact.* **35**, 12 (1995) 1693–1701.

22. A. Miller, The magic number seven, plus or minus two: Some limits on our capacity for processing information, *The Philosophical Review* **63** (1956) 81–87.

23. A. R. Barron, Predicted square error: A criterion for automatic model selection, *Self-Organizing Methods in Modeling: GMDH Type Algorithms*, ed. S. J. Farlow (Marcel-Dekker, New York, 1984).

24. N. Metropolis, A. Rosenbluth, M. Rosenbluth, A. Teller and E. Teller, Equation of state calculation by fast computing machines, *J. Chem. Physics* **21** (1953) 1087–1092.

25. Y. Chen, Y.-T. Im and J. Lee, Finite element simulation of solidification with momentum heat and species transport, *J. Material Processing Technology* **48** (1995) 571–579.

26. A. G. Gerber and A. C. Sousa, A parametric study of the Hazelett thin-salb casting process, *J. Material Processing Technology* **49** (1995) 45–56.

CHAPTER 4

COMPUTER TECHNIQUES AND APPLICATION OF PETRI NETS IN MECHANICAL ASSEMBLY, INTEGRATION, PLANNING, AND SCHEDULING IN MANUFACTURING SYSTEMS

AKIO INABA

Gifu Prefectural Research Institute of Manufactural Information Technology (Resident Office: Gifu Prefectural Research Institute of Industrial Products, 1288 Oze, Seki, Gifu 501-3265, Japan)

TASTUYA SUZUKI, SHIGERU OKUMA and FUMIHARU FUJIWARA

Graduate School of Engineering, Nagoya University, Furo-cho, Chikusa-ku, Nagoya, 464-8601, Japan

The examination of an assembly scheduling to improve productivity is becoming more common because of the progress of computer technology. Generally speaking, we must solve both a problem of planning for task sequence and a problem of resource allocation in the assembly scheduling. In most of the previous works, they have been considered independently with little interaction and been handled as a two-stage problem. In conventional scheduling problem (in the narrow sense), for example, Job Shop Scheduling has widely been discussed and only resource allocation problem has mainly been focused on it. On the other hand, in assembly planning problem, designer try to find the best assembly sequence based on the summation of costs assigned to each task *a priori*. However, the cost assigned to each task usually depends on the machine, which executes the allocated task, and also no concurrent operation is taken into account at planning stage. These facts tell us that two problems above mentioned have strong interaction with each other.

In this chapter, we propose a new scheduling method for assembly system in which sequence planning and machine allocation can be dealt with simultaneously based on a Timed Petri Net modeling. Our proposed search algorithm is originally based on A* algorithm which has widely been used in the field of AI. Since one of the drawback in applying A* may be lying in its calculation time, we also introduce several improvements to reduce the calculation time in this chapter. Firstly, we develop a guideline for decision of the estimate function in A* algorithm which plays an essential role in applying A* algorithm to assembly scheduling. Secondly, a new search method based on a combination of A* algorithm and supervisor technique is proposed. We also try to develop a search algorithm which can take into consideration the repetitive process in manufacturing system. Finally, we carry out some numerical experiments and verify that by introducing improvements in the above mentioned, the calculation time can drastically be reduced with little loss to the quality of solution.

Keywords: Scheduling; timed petri net; assembly; A* algorithm.

1. Introduction

The examination of an assembly scheduling to improve productivity is becoming more common due to the advancement of computer technology. Generally speaking, we must solve both a problem of planning for task sequence and a problem of resource allocation in the assembly scheduling. In most of the previous works, they have been considered independently with little interaction and been handled as a two-stage problem.

In conventional scheduling problem (in the narrow sense), for example, Job Shop Scheduling[1–4] has widely been discussed and only resource allocation problem has mainly been focused. On the other hand, in assembly planning problem, designer try to find the best assembly sequence based on the summation of costs assigned to each task *a priori*.[9] However, the cost assigned to each task usually depends on the machine which executes the allocated task and also no concurrent operation is taken into account at the planning stage.

These facts tell us that the above mentioned two problems have strong interaction with each other. Figure 1 shows an illustrative example on this point. The solution from viewpoint of planning is sequence (a), because the summation of costs is smaller than that of (b). However, a production time of sequence (b) can be smaller than that of sequence (a), when we are allowed to utilize machine $M1$ and $M2$ simultaneously. Figure 2 shows another example. The solution from viewpoint of

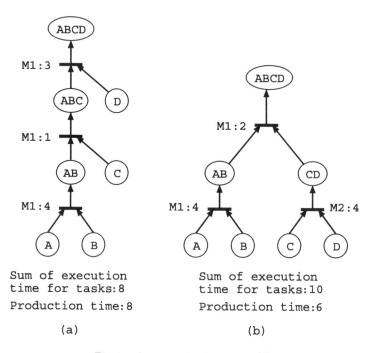

Fig. 1.   An example of planning (1).

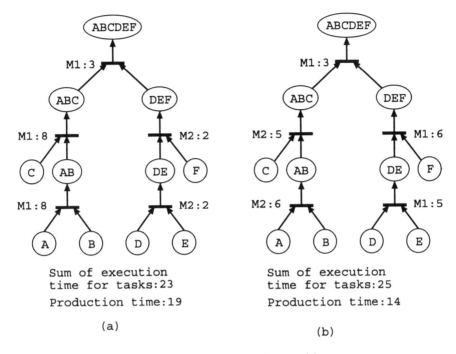

Fig. 2. An example of planning (2).

planning is sequence (a), but production time of sequence (b) can be much smaller than that of sequence (a), when we are allowed to utilize machine $M1$ and $M2$ simultaneously. (Here, M$\bigcirc$: $\bigcirc\bigcirc$ in Figs. 1 and 2 denote machine and execution time of each task in case of using assigned machine.) These considerations motivate us to develop the technique which enables us to handle the sequence planning and machine allocation simultaneously.

From this point of view, M. F. Sebaaly *et al.*[8] have proposed a simultaneous design based on genetic algorithm in which they restrict their attention to the case that the number of assembled product is only one for each kind of product. In many cases, however, the solution obtained by applying their approach does not always yield the best solution for the case that the number of product is greater than one.

In this chapter, we propose a new scheduling method for assembly system in which sequence planning and machine allocation can be dealt with simultaneously based on a Timed Petri Net modeling. Our proposed search algorithm is originally based on A^* algorithm which has widely been used in the field of AI. Since one of the drawback in applying A^* may be lying in its calculation time, we also introduce several improvements to reduce the calculation time in this chapter. Firstly, we develop a guideline for decision of the estimate function in A^* algorithm which plays an essential role in applying A^* algorithm to assembly scheduling. Secondly, a new search method based on combination of A^* algorithm and supervisor technique is proposed. We also try to develop a search algorithm which can take into consideration the repetitive process in manufacturing system. Finally, we carry out some numerical experiments and verify that by introducing improvements, the calculation time can drastically be reduced with little loss of the quality of solution.

Proposed method also has an advantage that the solution can be directly implemented on a Programmable Logic Controller because of the compatibility of the Petri Net and Programmable Logic Controller.

2. Modeling of Assembly Sequences

We create the model of assembly sequence including a manufacturing environment by doing stepwise refinement of assembly network which consists only of mechanical parts. Homem de Mellow et al.[9] have shown an efficient method for representation of assembly network using AND/OR graph. A corresponding Petri Net representation of assembly network is given by

$$N_S = (P_S, T_S, I_S, O_S), \tag{1}$$

where P_S, T_S, I_S and O_S are as follows.

$$P_S = \{\text{A set of subassemblies}\}$$
$$T_S = \{t_i\} = \{\text{A set of assembly tasks}\}$$
$${}^\bullet t_i = \{\text{two subassemblies before task } t_i\}$$
$$t_i^\bullet = \{\text{the subassembly after task } t_i\}$$
$$I_S(p_i, t_j) = \begin{cases} 1 : p_i \in {}^\bullet t_j \\ 0 : \text{otherwise} \end{cases}$$
$$O_S(t_i, p_j) = \begin{cases} 1 : p_j \in t_j^\bullet \\ 0 : \text{otherwise.} \end{cases}$$

Figure 4 shows Petri Net representation of assembly network about a ball-point pen shown in Fig. 3.

In Ref. 7, the authors have developed a technique to construct AND/OR Petri Nets from geometric information of mechanical product.

In this chapter, we assume that the assembly process is carried out under the manufacturing environment given as follows (Fig. 5).

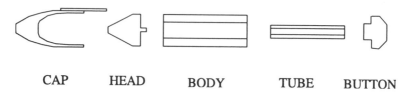

CAP   HEAD   BODY   TUBE   BUTTON

Fig. 3.   A ball-point pen.

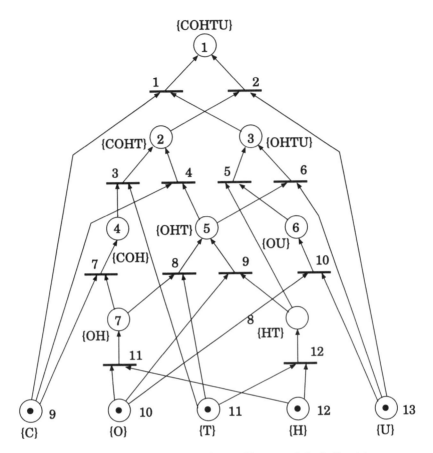

Fig. 4.   Petri net representation of assembly network for ball-point pen.

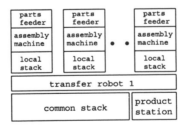

Fig. 5.   Production system.

- The production system consists of semi-universal assembly machines, a common stack, a product station and a transferring robot.
- Each assembly machine has its own parts feeder to supply it with parts.
- Each assembly machine has its own local stack and subassemblies are transferred between the machine and common stack through this local stack.
- The capacity of common stack is limited.

Under this manufacturing environment, each assembly task is supposed to be decomposed into the following steps.

- The transferring robot transfers subassemblies from the common stack to the local stack in specified assembly machine.
- The assembly machine assembles subassemblies on the local stack (or parts supplied from parts feeders).
- The transferring robot transfers the assembled subassembly from the local stack to the common stack.

Figure 6 shows Petri Net expression of these steps and its symbolic formulation is given as follows.

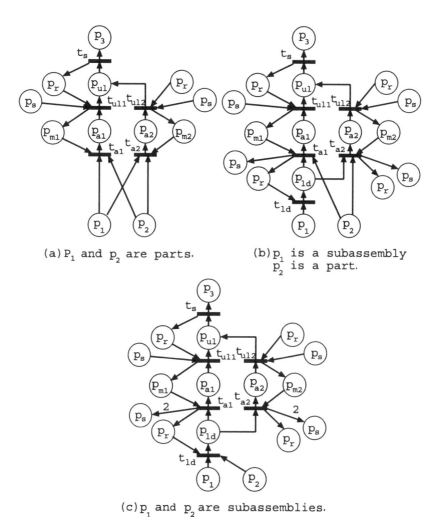

(a) P_1 and P_2 are parts.

(b) P_1 is a subassembly P_2 is a part.

(c) P_1 and P_2 are subassemblies.

Fig. 6.   Assembly tasks.

○ Set of places
● Set of places for subassemblies

$$P_{TS} = \{p_1, p_2, p_3\}, \tag{2}$$

where, p_1 and p_2 are subassemblies before assembling, p_3 is the assembled subassembly made from p_1 and p_2.
● Set of places for steps

$$P_{TT} = \begin{cases} \{p_{aj}, p_{ul} | j = 1, N_M\} & : p_1 \text{ and } p_2 \text{ are parts} \\ \{p_{ld}, p_{aj}, p_{ul} | j = 1, N_M\} & : \text{otherwise} \end{cases} \tag{3}$$

where p_{ld} is a step to load the assembly machine with subassemblies from the common stack. p_{aj} is a step to assemble subassemblies with machine j. p_{ul} is a step to unload subassemblies from the assembly machine to the common stack. N_M is number of machines in the environment.
● Set of places for resources

$$P_{TR} = \{p_r, p_s, p_{mj} : j = 1, N_M\} \tag{4}$$

where p_r is the transferring robot, p_s is the common stack, p_{mj} is the machine j for the assembly task. Each of these places has a token if it is available.

After all, the set of places for assembly task is given as follows.

$$P_T = P_{TS} \cup P_{TT} \cup P_{TR} \tag{5}$$

○ Set of transitions

$$T_T = \{t_j\} = \begin{cases} \{t_{ak}, t_{ulk}, t_s : k = 1, N_M\} & : p_1 \text{ and } p_2 \text{ are parts} \\ \{t_{ld}, t_{ak}, t_{ulk}, t_s : k = 1, N_M\} & : \text{otherwise} \end{cases} \tag{6}$$

where t_{ld}, t_{ak}, t_{ulk} and t_s are given as follows.
$^{\bullet}t_{ulk} = \{p_{ak}, p_r, p_s\}, t_{ulk}^{\bullet} = \{p_{ul}, p_{mk}\}$
$^{\bullet}t_s = \{p_{ul}\}, t_s^{\bullet} = \{p_3, p_r\}$
In the case that p_1 and p_2 are parts.
  $^{\bullet}t_{ak} = \{p_1, p_2, p_{mk}\}, t_{ak}^{\bullet} = \{p_{ak}\}$
In the case that $p_1(p_2)$ is a subassembly and $p_2(p_1)$ is a part.
  $^{\bullet}t_{ld} = \{p_1(p_2), p_r\}, t_{ld}^{\bullet} = \{p_{ld}\}$
  $^{\bullet}t_{ak} = \{p_2(p_1), p_{ld}, p_{mk}\}, t_{ak}^{\bullet} = \{p_{ak}, p_r, p_s\}$
In the case that p_1 and p_2 are subassemblies.
  $^{\bullet}t_{ld} = \{p_1, p_2, p_r\}, t_{ld}^{\bullet} = \{p_{ld}\}$
  $^{\bullet}t_{ak} = \{p_{ld}, p_{mk}\}, t_{ak}^{\bullet} = \{p_{ak}, p_r, p_s\}$

○ Weight of arcs

$$I_T(p_j, t_k) = \begin{cases} 1 : p_j \in {}^\bullet t_k \\ 0 : \text{otherwise} \end{cases} \tag{7}$$

$$O_T(t_j, p_k) = \begin{cases} 2 : p_k \in t_j^\bullet, p_k = p_s, p_1 \text{ and } p_2 \text{ are subassemblies} \\ 1 : p_k \in t_j^\bullet, (p_k \neq p_s \ or \ (p_1 \ or \ p_2 \text{ is a subassembly})) \\ 0 : \text{otherwise} \end{cases} \tag{8}$$

○ Set of time assigned to place

$$A_T = \{a_j\} \tag{9}$$

$$a_j = \begin{cases} \text{execution time} : p_j \in P_{TT} \\ 0 \qquad\qquad\qquad : \text{otherwise.} \end{cases} \tag{10}$$

After all, Petri Net corresponding to the assembly task is given by

$$N_T = (P_T, T_T, I_T, O_T, A_T). \tag{11}$$

We extend assembly network as shown in Fig. 4 to assembly system including manufacturing environment (resources) as follows (Fig. 7) . Here, we use subscript i, like N_T^i, to specify each task.

(1) We replace each transition in N_S by N_T^i.
(2) If successive assembly with one machine is possible, bypass nets are added because we do not need the unloading and reloading steps between assembly tasks.

Based on the above procedure, we consider the complete Petri Net representation for assembly system including resources. First of all, we give the bypass net to the assembly task p_{aj}^i executed with the machine j in the following manner.
○ In the case that one of p_1^i and p_2^i is a part
We add the transition which represents the beginning of successive assembly task for the subassembly p_1^i (p_2^i) with the machine p_{mj}^i, as shown in Fig. 7(a).

$$t_{aj}^{bi}(k, l) = \begin{cases} \{t_{aj-k}^{bi}\} : p_{mj}^i = p_{ml}^k, p_1^i(p_2^i) = p_3^k \\ \phi \qquad\quad : \text{otherwise} \end{cases} \tag{12}$$

where ${}^\bullet t_{aj-k}^{bi} = \{p_{al}^k, p_2^i\}(\{p_{al}^k, p_1^i\}), t_{aj-k}^{bi\bullet} = \{p_{aj}^i\}$.
• Sets of places and transitions along the bypass net

$$T_{pj}^i = \bigcup_{k=1}^{N_{TA}} \bigcup_{l=1}^{N_M^k} t_{aj}^{bi}(k, l) \tag{13}$$

$$P_{pj}^i = \phi \tag{14}$$

where N_{TA} is number of all tasks.

○ In the case that both p_1^i and p_2^i are subassemblies
We add the bypass net (bypass 1) which represents the process for successive

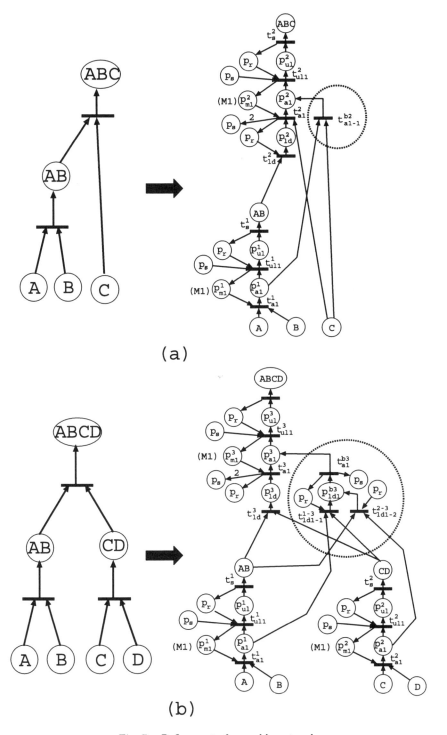

Fig. 7. Refinement of assembly network.

assembly task for the subassembly p_1^i in the machine p_{mj}^i, and add the bypass net (bypass 2) which represents the process of successive assembly task for the subassembly p_2^i with the machine p_{mj}^i (Fig. 7(b)). Here, we define the following transitions in the bypass net.

$$t_{ldj}^{1-i}(k,l) = \begin{cases} \{t_{ldj-k}^{1-i}\} : p_{mj}^i = p_{ml}^k, p_1^i = p_3^k \\ \phi \qquad : \text{otherwise} \end{cases} \tag{15}$$

$$t_{ldj}^{2-i}(k,l) = \begin{cases} \{t_{ldj-k}^{2-i}\} : p_{mj}^i = p_{ml}^k, p_2^i = p_3^k \\ \phi \qquad : \text{otherwise} \end{cases} \tag{16}$$

where $\bullet t_{ldj-k}^{1-i} = \{p_{al}^k, p_2^i, p_r\}$, $\bullet t_{ldj-k}^{2-i} = \{p_{al}^k, p_1^i, p_r\}$, $t_{ldk}^{1-i\bullet} = t_{ldk}^{2-i\bullet} = \{p_{ldj}^{bi}\}$ and p_{ldj}^{bi} is a step to load the local stack with subassembly from the common stack.
• Sets of places and transitions in bypass net

$$T_{pj}^i = \begin{cases} T_{pj}^{'i} \cup \{t_{aj}^{bi}\} : T_p^{'i} \neq \phi \\ \phi \qquad : \text{otherwise} \end{cases} \tag{17}$$

$$P_{pj}^i = \begin{cases} \{p_{ldj}^{bi}\} : T_p^{'i} \neq \phi \\ \phi \qquad : \text{otherwise} \end{cases} \tag{18}$$

where $T_{pj}^{'i} = \overset{N_{TA}}{\underset{k=1}{\cup}} \overset{N_M^k}{\underset{l=1}{\cup}} (t_{ldj}^{1-i}(k,l) \cup t_{ldj}^{2-i}(k,l))$, $\bullet t_{aj}^{bi} = \{p_{ldj}^{bi}\}$ and $t_{aj}^{bi\bullet} = \{p_{aj}^i, p_r, p_s\}$.

As the result, a complete Petri Net representation for an assembly system including resources is specified as follows.

$$N_A = (P_A, T_A, I_A, O_A, A_A), \tag{19}$$

where P_A, T_A, I_A, O_A and A_A are given as follows.

$$P_A = P_S \cup \left(\overset{N_{TA}}{\underset{i=1}{\cup}} \left(P_{TT}^i \cup P_{TR}^i \cup \left(\overset{N_M^i}{\underset{j=1}{\cup}} P_{pj}^i \right) \right) \right)$$

$$T_A = \overset{N_{TA}}{\underset{i=1}{\cup}} \left(T_T^i \cup \left(\overset{N_M^i}{\underset{j=1}{\cup}} T_{pj}^i \right) \right)$$

$$I_T(p_i, t_j) = \begin{cases} 1 : p_i \in \bullet t_j \\ 0 : \text{otherwise} \end{cases}$$

$$O_T(t_i, p_j) = \begin{cases} 2 : p_j \in t_i^\bullet, p_j = p_s, {}^\exists p_{ld}^k \in \bullet t_i, \\ \qquad p_{ld}^k = \{\text{task to load the local stack with two subassemblies} \\ \qquad\qquad \text{from the common stack.}\} \\ 1 : p_k^i \in t_j^{i\bullet}, (p_k^i \neq p_s \text{ or } {}^\exists p_{ld}^k \in \bullet t_i, \\ \qquad p_{ld}^k \neq \{\text{task to load the local stack with two subassemblies} \\ \qquad\qquad \text{from the common stack.}\} \\ 0 : \text{otherwise} \end{cases}$$

$$A_A = \{a_i\}$$

$$a_i = \begin{cases} \text{execution time} : p_i \in \bigcup_{i=1}^{N_{TA}} (P_{TT}^i \cup (\bigcup_{j=1}^{N_M^i} P_{pj}^i)) \\ 0 \qquad\qquad\qquad : \text{otherwise.} \end{cases}$$

The reason why we have adopted the Timed Place Petri Nets (TPPN) is that the marking in TPPN is always determined uniquely. On the contrary, if we use the Timed Transition Petri Nets (TTPN), tokens would disappear during the firing of transition, then the marking could not be determined uniquely.

3. Scheduling Algorithm

A scheduling algorithm executed based on the Petri Net model has been proposed by Doo Yong Lee *et al.*[5] This algorithm is a modified version of A^* algorithm which is widely used in the field of AI. The details of this algorithm can be described as follows.

Algorithm L1

STEP 1 Put the initial marking m_0 in the list $OPEN$.

STEP 2 If $OPEN$ is empty, terminate with failure.

STEP 3 Remove the first marking m from $OPEN$ and put it in the list $CLOSED$.

STEP 4 If m is final marking, construct the path from the initial marking to the final marking and terminate.

STEP 5 Find the enable transitions for the marking m.

STEP 6 Generate the next markings, or successor, for each enabled transition, and set pointers from the next markings to m. Compute $g(m')$ for every successor m'.

STEP 7 For every successor m' of m, do the following.

  a: If m' is already in $OPEN$, direct its pointer along the path yielding the smallest $g(m')$.

  b: If m' is already in $CLOSED$, direct its pointer along the path yielding the smallest $g(m')$. If m' requires pointer redirection, move m' to $OPEN$.

  c: If m' is not in either $OPEN$ or $CLOSED$, compute $h(m')$ and $f(m')$ and put m' in $OPEN$.

STEP 8 Reorder $OPEN$ in the order of magnitude of f.

STEP 9 Go to **STEP 2**.

Here, $f(m)$ is an estimate of the cost, i.e. the makespan from the initial marking to the final marking along an optimal path which goes through the marking m. $f(m) = g(m) + h(m)$. $g(m)$ is the current lowest cost from the initial marking to the current marking m. $h(m)$ is an estimate of the cost from the marking m to the final marking along an optimal path which goes through the marking m.

In this search algorithm, we can always find an optimal path if $h(\boldsymbol{m})$ satisfies the following condition.

$$h(\boldsymbol{m}) \leq h^*(\boldsymbol{m}) \quad \text{for all } \boldsymbol{m}, \tag{20}$$

where $h^*(\boldsymbol{m})$ is the actual cost of the optimal path from $\boldsymbol{m}$ to the final marking. The estimate function based on the number of remaining tasks was presented in literature.[5] However, no method to estimate the number of remaining tasks has been shown in it. In assembly problem, we can give the estimate as follows. We need $n-1$ tasks to assemble a product from n subassemblies. In this case, the number of remaining tasks is $n-1$. We extend this idea to the case that the number of products is α. At the intermediate state of assembly process, if the number of subassemblies are l, then the number of remaining tasks is $l - \alpha$. Therefore, we use the estimate function $h(\boldsymbol{m})$ defined as follows.

$$h(\boldsymbol{m}) = \frac{\sum_{i=1}^{k} f(i) \cdot \boldsymbol{m}(i) - \alpha}{N_M} \cdot C_{\min}, \tag{21}$$

where α, k and N_M are the number of products, places and machines respectively. $\boldsymbol{m}(i)$ is the number of token in place i at marking $\boldsymbol{m}$. $C_{\min}$ is the minimum cost in all assembly tasks. Also $f(i)$ is given as follows.

$$f(i) = \begin{cases} 1 : \text{place } p_i \text{ corresponds to a subassembly} \\ 0 : \text{otherwise.} \end{cases}$$

On the other hand, in algorithm L1, it is well known that when $h(\boldsymbol{m})$ is much smaller than $h^*(\boldsymbol{m})$, search time to find a solution becomes longer. Generally speaking, $h(\boldsymbol{m})$ shown in the Eq. (21) is much smaller than $h^*(\boldsymbol{m})$ because it does not take into account the cost of transferring task and few tasks may be performed with minimum cost $C_{\min}$. In order to take into account this point. We redefine $h(\boldsymbol{m})$ as follows.

$$h(\boldsymbol{m}) = \frac{\sum_{i=1}^{k} f(i) \cdot \boldsymbol{m}(i) - a}{N_M} \cdot (C_a + T_a), \tag{22}$$

where C_a is a mean cost of assembly tasks and T_a is a mean cost of transfer tasks.

Although Eq. (22) may not satisfy the condition (20), in many cases, the optimal solution can be obtained. The simulation results on this point will be shown in Sec. 7.

4. Reduction of Search Space Using a Supervisor

When we apply the algorithm L1 to a scheduling problem in which many numbers of products are assembled, search space becomes enormous and we need much time to solve it. In order to overcome this difficulty and obtain a quasi optimal solution with small calculation time, we propose a new algorithm which introduces the following three strategies.

(1) We add state feedback (a supervisor) to assembly Petri Nets in order to reduce the size of state space.

(2) We try to search a quasi optimal solution which includes repetitive operation.

(3) We impose a restriction on the capacity of list OPEN.

The strategy (3) is introduced in Ref. 6. In the remainder of this section, we state about (1). We will describe about (2) in the next section.

The supervisor controls a firing of each transition based on markings of assembly system and reduces the search space (Fig. 8). A control place is attached to each transition of assembly Petri Nets, and is linked to each transition by arcs (Fig. 9). The supervisor controls a existence of a token in the control place based on the state of marking in assembly Petri Nets and a closed loop specification. For example, when execution of a task is forbidden, the token in the control place connected to corresponding transition is removed. In the same way, firing of transitions can be controlled by the supervisor. On the other hand, this method can be viewed as a new search algorithm based on the combination of A^* algorithm and the supervisor.

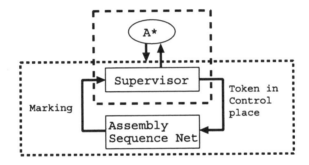

Fig. 8.   Search algorithm.

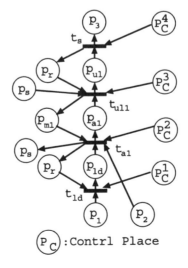

Fig. 9.   Control place.

In other words, we can say that a "rule based algorithm" (supervisor) is built in the original A^* algorithm in order to reduce the search space. In the following, we introduce a supervisor which handles a problem of parts allocation to parts feeders.

Here, we consider the following control specification.

• Same kind of part must be supplied from same parts feeder.

This is natural specification because, in practical manufacturing, there are few cases in which same kind of parts are supplied from different parts feeders. We show how to generate the control logic of supervisor for this specification.

4.1. *The control of allocation of parts feeders*

The supervisor controls tokens in the control places are based on Task List Table (Table 1) and Control List Table (Table 2). Task List Table includes information on the relationship among the machine number, supplied parts and places corresponding to the parts loading step. Control List Table includes information on the relationship among parts, loaded machine and transitions for loading task. For example, suppose that there is a token in place p_{a1}^1 at a marking m in an assembly system shown in Fig. 10, the supervisor detects that parts A and B are allocated to machine M_1 by referring Task List Table, and forbids the firing of transition for

Table 1.   Task list table.

Place	Machine No	Parts
p_{a1}^1	1	A, B
p_{a2}^1	2	A, B
p_{a1}^2	1	C, D
p_{a2}^2	2	C, D
p_{a1}^3	1	C
p_{a2}^3	2	C
p_{a1}^5	1	D

Table 2.   Control list table.

Parts	Machine No	Transitions for loading task
A	1	t_{a1}^1
A	2	t_{a2}^1
B	1	t_{a1}^1
B	2	t_{a2}^1
C	1	$t_{a1}^2, t_{a1}^3, t_{a1-1}^{b3}$
C	2	$t_{a2}^2, t_{a2}^3, t_{a2-1}^{b3}$
D	1	$t_{a1}^2, t_{a1}^5, t_{a1-3}^{b5}$
D	2	t_{a2}^2

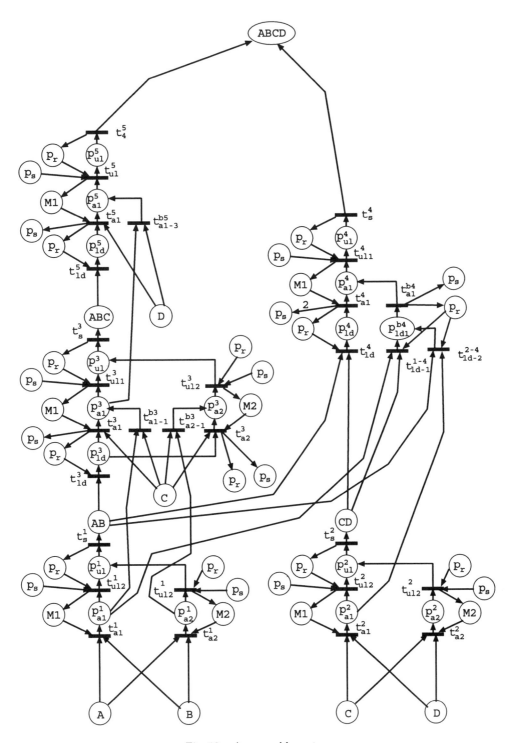

Fig. 10.   An assembly system.

steps corresponding to loading machines except for machine M_1 with parts A and B by referring Control List Table. In this case, transition t_{a2}^1 is forbidden.

4.2. *Avoidance of deadlock occurring from control of parts allocation*

In an assembly system shown in Fig. 10, when t_{a2}^2 fires and a token is provided in p_{a2}^2 (an assembling step for parts C and D by machine M_2 is performed), parts C and D are allocated to the parts feeder in machine M_2. Then, the supervisor introduced in previous subsection forbids firing of transitions t_{a1}^2, t_{a1}^5, t_{a1-3}^{b5}. This leads to the execution of assembly step for subassembly ABC and part D is impossible. However, when one of transitions t_{ld}^3, t_{a1-1}^{b3} and t_{a2-1}^{b3} is fired, it starts to make subassembly ABC and after all, results in deadlock. In order to avoid this type of deadlock, the supervisor has to forbid firing of transitions t_{ld}^3, t_{a1-1}^{b3} and t_{a2-1}^{b3}, when firing of transitions t_{a1}^5 and t_{a1-3}^{b5} are forbidden. Therefore, we introduce another tables called Subassembly List Table (Table 3) and Task Group List Table (Table 4). Subassembly List Table includes information on the relationship between subassembly s and group of assembly task including s. T_{out} is the group of task to assemble s and another subassembly or part. T_{in} is a group of tasks to make s. Task Group List Table includes information on the relationship among the task group, controlled transitions, entry transitions and entry places. The information on the controlled

Table 3.   Subassembly list table.

Subassembly	T_{in}	T_{out}
AB	$A + B$	$AB + CD, AB + C$
CD	$C + D$	$AB + CD$
ABC	$AB + C$	$ABC + D$

Table 4.   Task group list table.

Task Group	Controlled Transitions	Entry Transitions	Entry Places
$A + B$	$t_{a1}^1(p_{a1}^1)$ $t_{a2}^1(p_{a2}^1)$	t_{a1}^1 t_{a2}^1	p_{a1}^1, p_{a2}^1
$C + D$	$t_{a1}^2(p_{a1}^2)$ $t_{a2}^2(p_{a2}^2)$	t_{a1}^2 t_{a2}^2	p_{a1}^2, p_{a2}^2
$AB + C$	$t_{a1}^3(p_{a1}^3)$ $t_{a2}^3(p_{a2}^3)$	t_{ld}^3, t_{a1-1}^{b3} t_{a2-1}^{b3}	p_{ld}^3, p_{a1}^3 p_{a2}^3
$AB + CD$	$t_{a1}^4(p_{a1}^4)$	t_{ld}^4, t_{a1-1}^{1-4} t_{a1-2}^{2-4}	p_{ld}^4, p_{ld1}^{b4}
$ABC + D$	$t_{a1}^5(p_{a1}^5)$	t_{ld}^5, t_{a1-3}^5	p_{ld}^5, p_{a1}^5

transitions is used in order to detect forbidden tasks. Entry transitions express the beginning task in a task group and entry places are output places of entry transitions. Entry transition is used to forbid all steps in task group and entry place is used to detect the beginning of task in the task group. Based on these considerations, the supervisor controls tokens in control place as follows.

(1) When all tasks in T_{out} are disabled, forbid the execution of all tasks in T_{in}.
(2) When a task in T_{in} is executed, and only one task in T_{out} can be performed, disable the execution of all tasks of which execution disable it.
(3) If all tasks in task group are forbidden, disable the firing of all entry transitions of it.

For example, the supervisor introduced in previous section forbids t_{a1}^5 and t_{a1-3}^{b5}, when a token is in p_{a2}^2. The supervisor identifies that t_{a1}^5 for $ABC + D$ is forbidden by referring Task Group List Table, and forbid the execution of t_{ld}^5 by procedure (3). Moreover, the supervisor identifies that all tasks in T_{out} for subassembly ABC are forbidden by referring Subassembly List Table, then forbid $AB + C$ in T_{in} based on procedure (1) and forbids t_{ld}^3, t_{a1-1}^{b3} and t_{a2-1}^{b3} by referring Task Group List Table.

When t_{ld}^3 is fired and a token is provided into p_{ld}^3, the supervisor identifies the execution of task $AB + C$ by referring Task Group List Table. In this case supervisor executes procedure (2) because the supervisor identifies that only p_{a1}^5 belongs to T_{out} of subassembly ABC of which T_{in} includes $AB + C$ by referring Subassembly List Table. Assuming that there is a token in p_{a1}^5, the supervisor forbids t_{a2}^2 according to procedure in Sec. 4.1.

The mechanism of this deadlock avoidance does not always guarantee that all deadlock can be avoided because the supervisor uses only local information. However, the supervisor can remove many deadlocks and improve efficiency of search algorithm.

5. Speedup Based on Searching a Quasi Optimal Solution Including Repetitive Process

There are several cases that the optimal solution includes repetitive process in manufacturing system, when infinite number of products are assembled. In this chapter, paying attention to this point, we try to reduce the calculation time by searching a quasi optimal solution including repetitive process.

The repetitive process can be detected in the following manner. First, we define a marking time, a sojourn time of a token and an equivalent marking.

- **Marking time**
  Marking time is time when a transition becomes enable in marking.

- **Sojourn time of a token in a marking**
  Sojourn time of a token is the time interval from marking time to a time when the token comes to be able to go out. If sojourn time is negative, it is defined as 0.

- **Equivalent marking for m**

  Equivalent marking for m is a marking in which the number of token for each place except for places corresponding to parts and products is equal to that of m and also the sojourn time of each token in places except for places corresponding to parts and products is smaller than that of m or equal to that of m.

Figure 11 shows equivalence of markings. There is no place corresponding to parts and products in it. Each number assigned to place represents a time when the token comes to be able to go out. Each number in parentheses represents sojourn time. In Fig. 11(a), marking time is 10 because transition t_2 becomes enable at time 10. Sojourn time of the token in place p_1 is $15 - 10 = 5$. The marking (b) is equivalent marking for the marking (a) because number of token for each place in marking (b) is equal to that of marking (a) and sojourn time of each token for each place in marking (b) is smaller than that of marking (a) or equal to that of marking (a). The marking (c) is not equivalent marking for the marking (a) because sojourn time of token for place p_3 is longer than that of marking (a). The marking (d) is not equivalent marking for marking (a) either, because number of token for place p_2 in marking (d) is not equal to that of marking (a).

When a marking m at step 3 in algorithm L1 is compared with its parent marking and number of tokens in place corresponding to product changes, we start

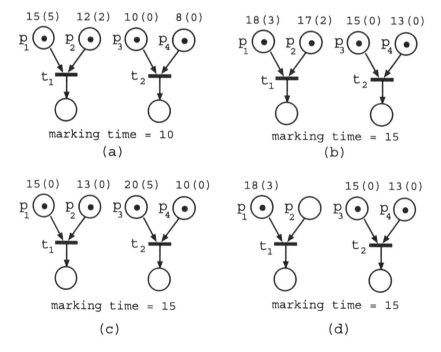

Fig. 11.   Equivalence of markings.

to examine the repetitive process. Repetitive process is detected by searching a marking m' satisfying following conditions in the list along the path from m to initial marking m_0.

- **Condition 1**
  m' is a equivalent marking of m.

- **Condition 2**
  Number of tokens in every place corresponding to products and parts at m' differs from one at m.

Here condition 1 guarantees that we can perform firing sequence from m to m' with same firing timing. Condition 2 expresses that the loading of all parts and the production of product are done. The reason why we detect repetitive process based on marking is that we do not have to consider a length of repetitive process.

A scheduling algorithm considering repetitive process is given as follows.

Algorithm L2

STEP 1 Put the initial marking m_0 in the list $OPEN$.

STEP 2 If $OPEN$ is empty, terminate with failure.

STEP 3 Remove the first marking m from $OPEN$ and put m in the list $CLOSED$.

STEP 4 When marking m satisfies following conditions, examine a repetitive process.

  a: Number of tokens in the place corresponding to the major product is equal to maximum number of tokens in the same place among all marking in $OPEN$.

  b: Number of tokens in each place corresponding to each product is different from one at the parent marking.

If repetitive process is detected, generate marking m^r by firing transitions which executes repetitive process. Set pointers from m^r to m. Compute $g(m^r)$ and put m^r in the list $CLOSED$. Regard m^r as m. Make the list $OPEN$ empty.

STEP 5 If m is final marking, construct the path from the initial marking to the final marking and terminate.

STEP 6 Find the enable transition at the marking m.

STEP 7 Generate the next marking, or successor, for each enabled transition, and set pointers from the next marking to m. Compute $g(m')$ for every successor m'. Control the tokens in control places by the supervisor.

STEP 8 For every successor m' of m, do the following.

  a: If m' is already in $OPEN$, direct its pointer along the path which yields the smallest $g(m')$.

b: If m' is already in $CLOSED$, direct its pointer along the path which yields the smallest $g(m')$. If m' requires pointer redirection, move m' to $OPEN$.

c: If m' is not in either $OPEN$ or $CLOSED$, compute $h(m')$ and $f(m')$ and put m' in $OPEN$.

STEP 9 Reorder $OPEN$ in order of the magnitude of f. If the number of elements in list $OPEN$ is greater than L_{open}, eliminate the markings of which value f is larger, so that the number of elements becomes smaller than L_{open}. Here, L_{open} is the capacity of list $OPEN$.

STEP 10 Go to **STEP 2**.

This algorithm does not guarantee that it can detect all repetitive operations especially when the number of products is small. However, if the number of products is very large, this algorithm could give us great advantage from the viewpoint of calculation time.

6. Examples

We show an example of assembly scheduling with the case of two assembly machines M_1 and M_2 and two kinds of products. Table 5 shows execution time of each task for products P_1 and P_2. P_1 and P_2 consist of four parts and their assembly network is given in Fig. 12. As shown in Table 5, the tasks 8, 9 and 10 are supposed to be executable by only M_1. Cost of transferring task is 1 per one subassembly. The characteristics of machines for each product is given as follows.

$P_1\mathbf{F}$ There is no explicit difference in the execution time between machines 1 and 2.
$P_2\mathbf{F}$ Execution time of M_1 is smaller than that of M_2 at each task.

We consider the case that the products P_1 and P_2 are assembled concurrently using same set of common stack, assembly machine and transferring robot. Figure 13 shows the solution for this scheduling problem. We can identify the effectiveness of scheduling considering both sequence planning and machine allocation concurrently. All tasks for making P_2 are allocated to M_1 because execution time of M_1 is smaller than that of M_2 at each task.

Table 5.   Production time of tasks.

Task	P_1	P_2	Task	P_1	P_2
1 ($M1$)	2	2	5 ($M2$)	4	4
1 ($M2$)	4	3	6 ($M1$)	3	4
2 ($M1$)	3	3	6 ($M2$)	4	5
2 ($M2$)	2	4	7 ($M1$)	3	4
3 ($M1$)	3	3	7 ($M2$)	4	5
3 ($M2$)	2	4	8 ($M1$)	3	3
4 ($M1$)	4	4	9 ($M1$)	3	3
4 ($M2$)	2	5	10 ($M1$)	2	2
5 ($M1$)	2	3			

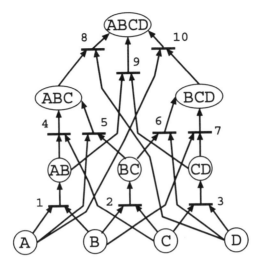

Fig. 12. An assembly network for a product.

7. Evaluation of Calculation Time

7.1. *Effect of estimate function*

In order to confirm the usefulness of the proposed estimate function, we compare the quality of solution and calculation time obtained from (22) with ones obtained from (21) in case of assembling two products of which assembly network is given by Fig. 12. Costs of assembly step in the products are supposed to take random values which vary from 2 to 7. Cost of transferring task is 1 per one subassembly. Calculation times listed on Table 6 are average time of 50 trials. The computer used for calculation is CELEBRIS GL 6200 (pentium pro 200 MHz).

In Table 6, we cannot identify difference in the quality of solution between (21) and (22). However, the calculation time using (22) is much smaller than one using (21). Thus, we can verify the usefulness of (22).

7.2. *Effect of searching repetitive process*

In order to confirm the effectiveness of detecting the repetitive process, we compare the quality of solution and calculation time in the case of taking into account the repetitive process with those not considering the repetitive process. We examine the case of assembling 30 products of which assembly network is given by Fig. 12. Costs of assembly tasks in the products take random value which vary from 2 to 7. Cost of transferring task is 1 per one subassembly. Limitation of capacity for list OPEN is set to be 200 and the supervisor is used in both cases. Calculation times listed on Table 7 are an average time of 30 trials in which we can calculate solutions within two hours.

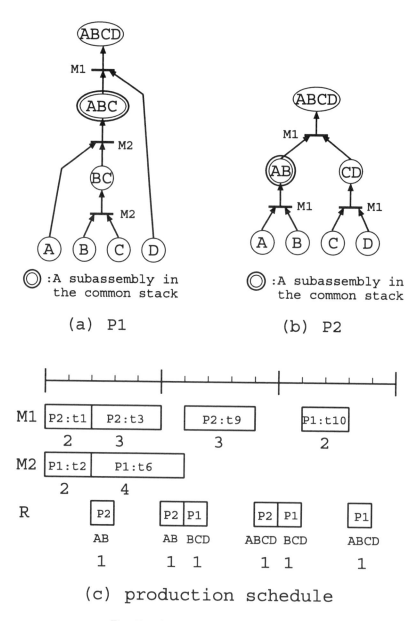

Fig. 13.   Assemble schedule of example.

As shown in Table 7, we cannot identify difference in the quality of solution between two cases. However, the calculation time in the case of considering the repetitive process is much smaller than that of normal search. Thus, the usefulness of considering the repetitive process is verified.

Table 6.   Execution time (1).

	Eq. (21)	Eq. (22)
Production time	15.1	15.4
Execution time	12.2 s	3.9 s

Table 7.   Execution time (2).

	Consideration of Repetitive Process	Without Consideration of Repetitive Process
Production time	215.3	215.2
Execution time	33.9 s	1108.6 s

8. Conclusions

In this chapter, we have proposed a new scheduling method in which sequence planning and resource allocation are considered simultaneously. First of all, we have proposed a modeling method for an assembly sequence including a manufacturing environment with timed Petri Nets. By using the model based approach, we can always obtain feasible sequences without any consideration. Secondly, we have developed a guideline for decision of the estimate function in A^* algorithm which plays an essential role in applying A^* algorithm to assembly scheduling. Thirdly, a new search method based on combination of A^* algorithm and the supervisor has been proposed. We have also developed a search algorithm which can take into consideration the repetitive process in manufacturing system. Finally, we have carried out some numerical experiments and verified that by introducing improvements above mentioned, the calculation time has been drastically reduced with little loss of the quality of solution.

References

1. S. M. Johnson, Optimal two-and-three-stage production schedules with setup times included, *Nav. Res. Log. Quart.* **1**, 1 (1954) 61–68.
2. J. R. Jackson, An extension of Johnson's results on job-lot scheduling, *Nav. Res. Log. Quart.* **3**, 3 (1956) 201–203.
3. J. H. Blackstone, D. T. Phillips and G. L. Hogg, A state-of-the-art survey of dispatching rules for manufacturing job shop operations, *Int. J. Production Research* **20**, 1 (1982) 27–45.
4. K. Tanaka and N. Ishii, Scheduling and simulation, *SICE*, 1955 (in Japanese).
5. D. Y. Lee and F. DiCesare, Petri net-based heuristic scheduling for flexible manufacturing, *Petri Nets in Flexible and Agile Automation*, ed. Mong Chu Zhou (Kluwer Academic Publishers, Boston, 1995) 149–188.
6. D. Y. Lee and F. DiCesare, Scheduling flexible manufacturing systems using petri nets and heuristic search, *IEEE Transactions on Robotics and Automation* **10**, 2 (1994) 123–132.

7. A. Inaba, T. Suzuki and S. Okuma, Generation of assembly or disassembly sequences based on topological operations, *Transactions of the Japan Society of Mechanical Engineers(C)* **63**, 609 (1997) 1795–1802.

8. M. F. Sebaaly and H. Fujimoto, Integrated planning and scheduling for multi-product job-shop assembly based on genetic algorithms, *Proceedigns of the 6th IFIP TC5/WG5.7 International Conference on Advances in Production Management Systems — APMS'96*, Kyoto, Japan, 1996, pp. 557–562.

9. L. S. Homen de Mello and A. C. Sanderson, AND/OR graph representation of assembly plans, *IEEE Transactions on Robotics and Automation* **6**, 2 (1990) 188–199.

CHAPTER 5

DATA AND ASSEMBLY TECHNIQUES AND THEIR APPLICATIONS IN AUTOMATED WORKPIECE CLASSIFICATION SYSTEM

S. H. HSU* and M. C. WU

Department of Industrial Engineering and Management,
National Chiao Tung University,
1001, Ta-Hsueh Road,
Hsinchu, Taiwan, Republic of China
* shhsu@cc.nctu.edu.tw; Fax: 886-3-5722392

T. C. HSIA

Department of Industrial Engineering and Management,
Chien Kuo Institute of Technology,
1, Ga-Shou N. Road,
Changhua, Taiwan, Republic of China

In group technology, workpieces are categorized into families according to their similarities in designs or manufacturing attributes. This technique can eliminate design duplication and facilitate production. Much effort has been focused on automated workpiece classification systems development. However, it is difficult to evaluate the performance of these systems. In this study, a benchmark classification system was developed based on data and assembly techniques with global shape information to evaluate the utility of workpiece classification systems. A classification system has a high level of utility if its classification scheme is consistent with the user's mental model of the similarity between workpiece shapes. Hence, in the proposed method, the consistency between a classification system and the user's mental model is used as an index for the utility of the system. The proposed benchmark classification developed from the data and assembly techniques has two salient characteristics: (1) it is user-oriented, because it is based on users' judgments; (2) it is flexible, allowing users to adjust the criteria of similarity applied in the automated workpiece classification.

Such benchmark classification system is typically established by having subjects to perform complete pair comparisons of all sample workpieces. However, when the number of sample workpieces is very large, such exhaustive comparisons become impractical. The authors also propose an efficient method, called lean classification system, in which data on comparisons between the samples and a small number of typical workpieces are used to infer the benchmark classification system results.

Keywords: Automated workpiece classification systems; benchmark classification; workplace classification; distribution patterns of workpieces.

1. Introduction

Workpiece coding schemes are widely used in the implementation of group technology (GT) to classify workpieces according to the similarity of their designs and manufacturing attributes. The results of workpiece classification can be used to establish design and manufacturing databases, which facilitate the retrieval of similar designs and the standardization of manufacturing processes and thus enhance design and manufacturing productivity.

The design and manufacturing attributes used in coding workpieces generally involve shape (i.e. geometric form and size), function, material, and other manufacturing characteristics. In recent years, shape-related attributes have drawn much attention from researchers because of the increasing demand for fully integrated CAD/CAM systems.

Manual coding of workpieces on the basis of their shape is a time-consuming and error-prone process. The operator has to memorize all of the template-shapes and then match a particular template-shape with each workpiece. Few operators can perform such matching accurately and reliably, especially when a large number of workpieces are involved. To overcome this problem, researchers have developed several automated classification systems such as Bhadra and Fischer,[3] Chen,[6] Henderson and Musti,[10] and Kaparthi and Suresh.[13] Most of these approaches use individual local geometric features as the descriptors for workpiece classification, and approaches of this type have been shown to be useful in the planning of manufacturing processes.

However, there are two shortcomings in using individual features as classification criteria:

(1) As Fig. 1 shows, similarity of isolated individual features does not necessarily entail similarity in global shape.
(2) Isolated individual features cannot be used for identification during the early stages of the design phase, because the designer's conceptual model evolves from an overall, global picture to individual details.

Thus, the use of local features as workpiece classification criteria is generally limited to information retrieval and practical applications.

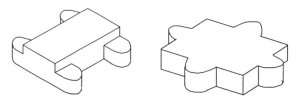

Fig. 1. Two workpieces with similar individual features but significantly different in global shape information.

In more recent GT research, workpieces have been described and classified on the basis of the overall contour of the workpiece instead of local attributes such as Lenau and Mu,[17] Wu and Chen,[21] Wu *et al.*,[22] and Wu and Jen.[23] This approach enhances performance in the design phase and increases efficiency on the manufacturing and assembly lines. One of the key criteria in choosing a practical automated classification system for design, manufacturing, and assembly is whether the classification results are compatible with the user's own classification. To ensure that this criterion is met, benchmarks reflecting the user's classification are needed to evaluate the performance of automated classification systems.

There is, however, no research regarding whether the classification result meets the user requirement although the execution speed of such an automated classification is remarkably high. The evaluation of the utility of workpiece classification system is a must in choosing a proper workpiece classification system that is compatible to the user's cognition in designing, manufacturing, and assembling. The utility of a classification system means that it is easy for users to store and search the database. Moreover, the ease of workpiece storage and retrieval lies in the compatibility of the framework of the classification system and the user's mental model.

The first objective of this study is to develop a benchmark classification system based upon the user's cognitive structure through the data and assembly techniques in order to evaluate the utility of an automated workpiece classification system.

To establish such classification benchmarks, Hsu *et al.*[11] presented a benchmark classification technique in which sets of sample workpieces were selected from the general population of workpieces. Pair comparisons of the sample workpieces on the basis of the similarity of the global shape of the workpieces were then made. The degree of similarity between each pair of sample workpieces obtained from the overall results of the user comparisons was called the complete experimental data. These data were then used to classify the samples. The classification results were regarded as a benchmark classification for the sample workpieces. By comparing these classification results with those of various automatic classification systems for the same set of samples, one can identify an automated classification system that produces results most consistent with users' judgments. Yet benchmark classification is also problematic, because it requires exhaustive pair comparisons of all of the sample workpieces. If the number of workpieces is n, the number of comparisons that must be made is at least $n(n-1)/2$. That is, the testing needed to classify a large set of sample workpieces will be extremely time-consuming. Moreover, when experimental subjects are required to make a very large number of pair comparisons, they are likely to become fatigued and produce biased or inconsistent judgments.

To reduce the manpower and time for workpiece benchmark classification, a second objective is derived. Namely, it is necessary to investigate whether or not the partial experimental data can take the place of the complete experimental data in building a lean classification system. If so, the partial experimental data can function as the classification specification criteria. Such an idea takes advantage of the partial experimental data and an inference technique to develop quasi-complete experimental data just like the complete experimental data for the pair comparison of all sample workpieces. This classification is defined as a lean classification. If there is a high degree of consistency between lean classification and benchmark classification, then lean classification can substitute for benchmark classification.

For the first objective, a benchmark classification system based upon the user's cognitive structure will be set up to evaluate the utility of the workpiece classification system. This method employs fuzzy clustering analysis to process the pair comparison of the target sampling workpiece in terms of their similarities. The cluster obtained from this process is called the benchmark classification. Furthermore, the classification of the same sample workpiece using the automated classification is defined as the test classification. Finally, the efficiency of the automated classification system can be examined by analyzing the difference between the benchmark classification and the test classification. In this way, the most practical automated workpiece classification can be selected.

Because the outcome of classification obtained from the fuzzy clustering analysis varies at different levels in the hierarchy, the user can choose the proper clustering as the evaluation basis according to their knowledge of the workpiece similarities. The benchmark classification, therefore, features the flexibility in application to the practical cases.

For example, if a user adopts a stringent criterion, so that only workpieces with a high degree of similarity are classified as belonging to the same group, then a benchmark classification will be obtained. This classification includes more groups, with each containing a small number of workpieces. On the other hand, if the user decides that for a particular application, even workpieces with a low degree of similarity can be grouped together, then a benchmark classification will be obtained that includes fewer groups, but each group will contain more workpieces. Since the criteria for forming the benchmark classification can be adjusted freely, this kind of classification process can be referred to a flexible classification method.

After the benchmark classification is determined, two indices are used to measure the level of consistency between the test classification and the benchmark classification. The first is an index of the appropriate number of workpiece groups, which is used to check whether there are too many or too few groups. If there are too many groups, then the similarity criteria used in the test classification system was too stringent. On the other hand, if there are too few groups, then the similarity criteria is too loose. The second index is the degree of consistency between two classifications. This index can calculate the ratio of the sum of the number of workpieces in each corresponding group in the test classification and the benchmark classification that are the same over the total number of workpieces within each group. A one-to-one correspondence is found between groups in the benchmark classification and the test classification by working from the classification with fewer groups. For instance, if the benchmark classification has fewer groups, then a correspondence is assigned between the groups in the benchmark classification and the groups in the test classification that are the most similar. The higher the level of similarity between each pair of corresponding groups, the more accurate the test classification.

As far as the second objective is concerned, it is necessary to verify the feasibility of using the lean classification where partial experimental data are employed to replace the benchmark classification. In the lean classification method, partial experimental data will be selected from the complete experimental data. This data will then be manipulated through a max–min method of the fuzzy set theory to infer the other data not chosen in the beginning. The aggregation of these two sets of data will bring forth the quasi-complete experimental data. With the fuzzy clustering technique analysis, the quasi-complete experimental data will be classified. Such a process is defined as the lean classification. Furthermore, the specific algorithm developed in this study will calculate the consistency and effectiveness of the lean classification system. If the consistency between the lean classification and the benchmark classification systems and if the lean classification system meets the utility requirement, then it will be appropriate for the user to use the lean classification rather than the benchmark classification system.

This paper's content is organized into seven sections. Section 1 presents the conceptual foundation of the automated workpiece classification system. In Sec. 2, empirical experiments are used to establish the benchmark classification system. Section 3 develops an index of the appropriate number of workpiece groups and the degree of consistency between the two classifications as the evaluation criteria to verify the utility of the automated workpiece classification system. Section 4 is aimed at establishing the lean classification system using partial experimental data. Section 5 measures the effectiveness and efficiency of lean classification through the degree of consistency between the lean classification system and the benchmark classification system. In Sec. 6, a large sampling of workpieces (100 to 800 items) is simulated to evaluate whether or not the lean classification system is effective and efficient. Finally, the authors arrive at some conclusions and offer suggestions for future work in Sec. 7.

2. Experiments to Establish the Benchmark Classification System

In this section, we describe an experimental study that was conducted to carry out workpiece classification. First of all, a set of sample workpieces was chosen from the general workpiece population. In an automatic workpiece classification system, if a large number of workpieces must be classified, there are likely to be a large number of workpiece categories. In this research, a comparatively small number of workpieces and the similarity of these sample workpieces were compared manually to establish a classification benchmark. Sample workpieces were selected according to a stratified sampling method.[20] A broad estimate was made of the number of workpiece categories likely to be appropriate in a practical environment in which an automatic classification system is used. The number of workpieces in each category was estimated, and then sample workpieces were selected from the general population according to the proportion of the number of workpieces in each category. For instance, suppose there are nine categories of workpieces in a certain design and manufacturing environment and there are approximately the same number of workpieces in each of these nine categories. The total number of sample workpieces will be 36 (see Fig. 2) if we randomly select four workpieces from each category. (For convenience, this set of 36 sample workpieces will be used as an example for the remainder of the paper.)

Secondly, we can use these sample workpieces to establish the benchmark classification system. The process of establishing the benchmark workpiece classification can be divided into two stages: (1) Using sample workpieces to get the complete experimental data; and (2) Establishing the benchmark classifications, as shown in Fig. 3.

2.1. Using Sample Workpieces to Get the Complete Experimental Data

The aim of this stage is to establish the benchmark workpiece classification. In this research, 30 subjects were asked to make pair comparisons of the global shape of the 36 sample workpieces shown in Fig. 2 to get the complete experimental data. When subjects are asked to compare the similarity of various objects, their judgements are often limited by their attention span, memory capacity, and previous experience. Their responses were generally fuzzy and could not be expressed using crisp

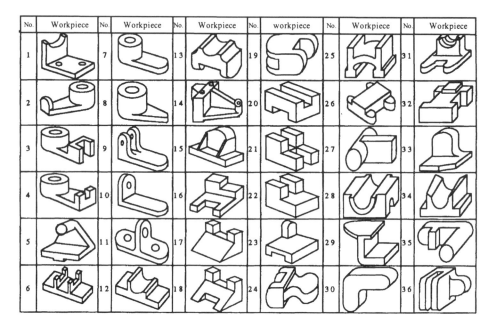

Fig. 2.   The 36 sample workpieces used for pair comparison.

1. Using sample workpieces to get the complete experimental data
(1) Selection of subjects (2) Method of representing workpieces for comparison (3) Definition of linguistic terms and membership functions (4) Aggregation of membership functions and defuzzification

2. Establishing the benchmark classifications
(5) Fuzzy clustering analysis (6) Using aggregated comparison crisp numbers for workpieces clustering

Fig. 3.   The benchmark workpiece classification process can be divided into two stages.

numbers from 0 to 9. To cope with the fuzziness inherent in the human cognitive processes, we used linguistic variables for the similarity comparison and integrated the subjects' responses by means of a fuzzy number operation. The fuzzy numbers were then changed into crisp numbers through a defuzzification process. These crisp numbers are called the complete experimental data. The detailed procedures are described below.

2.1.1. *Selection of subjects*

Since the aim of this project was to employ the users' intuitive classification of workpieces as a benchmark classification for evaluating the performance of automated classification systems, it was important that the subjects be truly representative of the intended users of such systems. Ideally, in a project of this type, the subjects should be randomly and proportionally selected from all departments of the factory that will use the automatic classification system. In this case, 30 subjects were selected from various departments of an aircraft plant. The sample was designed to reflect the actual proportions of the firm's workers employed in various tasks: 22 of the subjects were operators who worked directly on the production line (5 in the tooling design section and the other 17 in manufacturing and structural assembly), and 8 were employed in program management, production control, and other departments.

2.1.2. *Method of representing workpieces for comparison*

Workpieces are 3D objects. There are two methods of representing a 3D object in a 2D drawing. First, the object can be represented by means of three 2D orthogonal drawings (front view, side view, and top view) as shown in Fig. 4(a). This method depicts the shape of the object exactly, but makes it difficult for the subject to

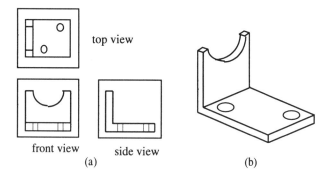

Fig. 4. Workpieces represented in (a) 2D orthogonal drawing and (b) isometric drawing.

picture the entire 3D object mentally. Secondly, the object can be presented by means of an isometric drawing, as shown in Fig. 4(b). This type of drawing conveniently depicts both the overall structure and individual details of the 3D object, and because the drawing conveys a great deal of information in a compact form, it facilitates recognition and comparison of objects.[8] However, because isometric drawings provide only one view of an object, they can be misleading in at least two ways. First, two objects with a similar shape may appear more alike than they really are, because their differentiating features may be hidden from a certain angle of view. Secondly, as mentioned by Arnhiem,[1] an object presented in a 2D drawing will carry more visual weight in the upper or left part of the drawing. To prevent these two facts from biasing the subjects' judgement of the similarity of the sample workpieces, in this research the subjects were presented with front, back, top, bottom, left, and right isometric views of each of the workpieces, as shown in Fig. 5. In this way the subjects could easily determine the overall shape of the workpieces, and they then selected linguistic terms on an answer sheet to indicate the level of similarity between various workpieces. The evaluation scale used offered a choice of five terms: very low similarity, low similarity, medium similarity, high similarity, and very high similarity.

To compare all possible pairs of the 36 sample workpieces, the subjects had to make a total of 630, or C_2^{36}, individual comparisons. In order to ensure that the subjects could make consistent judgements, each subject worked independently, without a time limit. The 630 answer sheets were randomly distributed to the subjects. The results were coded by using the letters $\{A, E, I, O, U\}$ to denote the linguistic terms {very high similarity, high similarity, medium similarity, low similarity, very low similarity}, as shown in Table 1.

2.1.3. Definition of linguistic terms and their membership functions

As proposed by Chen and Hwang,[7] the membership function for each of the five linguistic terms was taken to be a trapezoidal fuzzy number, as shown in Fig. 6 (see the Appendix).

Fig. 5. Workpieces represented in 6 isometric views for comparison. U, very low similarity; O, low similarity; I, medium similarity; E, high similarity; A, very high similarity.

2.1.4. *Aggregation of membership functions and defuzzification*

In this step of the analysis, fuzzy number operations were employed to aggregate the information collected in the preceding stages. The membership function for each linguistic term can be seen as a fuzzy number. Linguistic terms chosen by more than two subjects were integrated into membership functions through operations with fuzzy numbers. There are many methods of aggregating a decision-maker's fuzzy assessments, such as mean, median, maximum, minimum, and mixed operators.[5] The most commonly used aggregation method, however, is the average operation. In the present research, the average operator was used to aggregate the subjects' judgements of the similarity of various pairs of workpieces using the following formula:

$$\tilde{S}_{ij} = \left(\frac{1}{n}\right) \otimes (\tilde{S}_{ij1} \oplus \tilde{S}_{ij2} \oplus \ldots \oplus \tilde{S}_{ijk} \oplus \ldots \oplus \tilde{S}_{ijn}) \qquad (1)$$

where
$\oplus$ represents addition of fuzzy numbers
$\otimes$ represents multiplication of fuzzy numbers

S. H. Hsu et al.

Table 1. The fuzzy relationship of 36 workpieces by pair comparison.

Department: Tooling design section of engineering department

Subject: × × ×

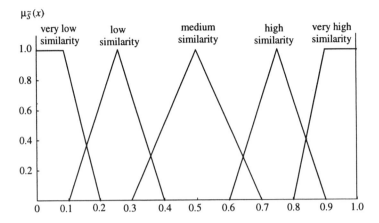

Fig. 6.   Membership functions for linguistic terms indicating level of similarity.

$\tilde{S}_{ijk}$ represents the membership function for the linguistic term, which is obtained by taking the number of subjects k comparing the similarity of workpiece i and workpiece j, where $k = 1, \ldots, n, 1 \le i < j \le w, n = 30$ (number of subjects), and $w = 36$ (number of workpieces)

$\tilde{S}_{ij}$ represents the membership function after aggregation of the n subjects' similarity comparison between workpiece i and workpiece j, where $i < j$.

In order to perform a crisp number calculation of the aggregated membership functions, it is necessary to determine a crisp value for the fuzzy sets concerned. This type of a transition process is called defuzzification, which generally uses the center of gravity or α-cut element average method.[18] In this research, the center of gravity method was used to seek the solutions, because, in terms of geometry, the center of gravity is the most representative point of the fuzzy set. The aggregated membership function is expressed in Eq. (2), and its value is denoted by X_G.[4]

$$X_G(\tilde{S}) = \frac{\int_0^1 x\mu_{\tilde{S}}(x)dx}{\int_0^1 \mu_{\tilde{S}}(x)dx}. \tag{2}$$

Because the linguistic term chosen by the subject to express the degree of similarity between workpieces is a trapezoidal fuzzy number, the aggregated membership function remained a trapezoidal fuzzy number. If the defuzzification with the center of gravity formula (2) is simplified, so that it is denoted by a trapezoidal fuzzy number (a, b, c, d), it is easy to determine a crisp comparison number for the fuzzy relationships among the aggregated 36 sample workpieces. These crisp numbers are called the complete experimental data and denoted as $\tilde{R}$ (Table 2). The simplified formula can be expressed as follows:

$$X_G(\tilde{S}) = \frac{-a^2 - b^2 + c^2 + d^2 - ab + cd}{3(-a - b + c + d)}. \tag{3}$$

Table 2.   The crisp numbers for the fuzzy relations among 36 sample workpieces, $\tilde{R}$.

	1	2	3	4	5	6	7	8	9	10	11	12	13	14	15	16	17	18
1	1.000	0.403	0.356	0.467	0.250	0.464	0.428	0.300	0.489	0.650	0.558	0.425	0.236	0.294	0.275	0.389	0.339	0.369
2	0.403	1.000	0.742	0.789	0.389	0.225	0.792	0.762	0.508	0.581	0.428	0.241	0.316	0.383	0.306	0.188	0.136	0.145
3	0.356	0.742	1.000	0.823	0.408	0.286	0.675	0.667	0.397	0.411	0.403	0.250	0.252	0.361	0.328	0.336	0.302	0.241
4	0.467	0.789	0.823	1.000	0.494	0.336	0.759	0.725	0.483	0.483	0.428	0.250	0.247	0.417	0.300	0.408	0.344	0.303
5	0.250	0.389	0.408	0.494	1.000	0.280	0.397	0.397	0.225	0.250	0.230	0.286	0.250	0.667	0.403	0.275	0.289	0.294
6	0.464	0.225	0.286	0.336	0.280	1.000	0.179	0.162	0.350	0.314	0.322	0.642	0.378	0.185	0.291	0.597	0.425	0.433
7	0.428	0.792	0.675	0.759	0.397	0.179	1.000	0.818	0.622	0.667	0.533	0.239	0.322	0.380	0.308	0.222	0.131	0.142
8	0.300	0.762	0.667	0.725	0.397	0.162	0.818	1.000	0.508	0.539	0.442	0.191	0.247	0.389	0.331	0.156	0.151	0.145
9	0.489	0.508	0.397	0.483	0.225	0.350	0.622	0.508	1.000	0.798	0.736	0.330	0.394	0.264	0.339	0.241	0.215	0.221
10	0.650	0.581	0.411	0.483	0.250	0.314	0.667	0.539	0.798	1.000	0.781	0.291	0.314	0.255	0.397	0.260	0.172	0.184
11	0.558	0.428	0.403	0.428	0.230	0.322	0.533	0.442	0.736	0.781	1.000	0.319	0.286	0.255	0.325	0.156	0.095	0.147
12	0.425	0.241	0.250	0.250	0.286	0.642	0.239	0.191	0.330	0.291	0.319	1.000	0.569	0.199	0.411	0.342	0.330	0.339
13	0.236	0.316	0.252	0.247	0.250	0.378	0.322	0.247	0.394	0.314	0.286	0.569	1.000	0.190	0.447	0.266	0.238	0.249
14	0.294	0.383	0.361	0.417	0.667	0.185	0.380	0.389	0.264	0.255	0.255	0.199	0.190	1.000	0.408	0.247	0.257	0.280
15	0.275	0.306	0.328	0.300	0.403	0.291	0.308	0.331	0.339	0.397	0.325	0.411	0.447	0.408	1.000	0.367	0.422	0.444
16	0.389	0.188	0.336	0.408	0.275	0.597	0.222	0.156	0.241	0.260	0.156	0.342	0.266	0.247	0.367	1.000	0.589	0.639
17	0.339	0.136	0.302	0.344	0.289	0.425	0.131	0.151	0.215	0.172	0.095	0.330	0.238	0.257	0.422	0.589	1.000	0.864
18	0.369	0.145	0.241	0.303	0.294	0.433	0.142	0.145	0.221	0.184	0.147	0.339	0.249	0.280	0.444	0.639	0.864	1.000
19	0.165	0.350	0.369	0.272	0.199	0.202	0.336	0.297	0.467	0.408	0.389	0.286	0.684	0.182	0.378	0.300	0.255	0.289
20	0.272	0.162	0.381	0.303	0.241	0.561	0.142	0.107	0.216	0.185	0.151	0.375	0.405	0.181	0.272	0.600	0.500	0.525
21	0.325	0.162	0.358	0.316	0.252	0.575	0.170	0.153	0.249	0.227	0.142	0.333	0.303	0.269	0.291	0.742	0.569	0.653
22	0.397	0.216	0.414	0.406	0.258	0.608	0.204	0.153	0.333	0.269	0.173	0.392	0.328	0.289	0.305	0.753	0.633	0.661
23	0.325	0.322	0.314	0.283	0.317	0.369	0.261	0.250	0.331	0.400	0.331	0.367	0.386	0.286	0.558	0.469	0.350	0.383
24	0.221	0.406	0.367	0.361	0.250	0.224	0.392	0.344	0.556	0.444	0.436	0.327	0.628	0.239	0.453	0.247	0.227	0.252
25	0.286	0.204	0.202	0.213	0.213	0.381	0.170	0.133	0.182	0.193	0.170	0.531	0.422	0.174	0.205	0.303	0.275	0.325
26	0.250	0.303	0.222	0.244	0.241	0.428	0.283	0.227	0.328	0.330	0.381	0.353	0.275	0.176	0.367	0.258	0.241	0.252
27	0.289	0.542	0.517	0.572	0.461	0.185	0.642	0.575	0.367	0.436	0.431	0.225	0.269	0.353	0.336	0.199	0.188	0.176
28	0.327	0.208	0.221	0.199	0.225	0.366	0.191	0.185	0.168	0.167	0.207	0.575	0.453	0.199	0.272	0.258	0.236	0.255
29	0.289	0.311	0.258	0.261	0.436	0.238	0.241	0.367	0.264	0.319	0.286	0.417	0.358	0.252	0.367	0.297	0.308	0.303
30	0.380	0.280	0.210	0.207	0.244	0.151	0.378	0.347	0.364	0.558	0.450	0.224	0.286	0.224	0.311	0.244	0.247	0.207
31	0.536	0.333	0.389	0.389	0.297	0.347	0.353	0.342	0.461	0.483	0.417	0.403	0.411	0.339	0.517	0.531	0.394	0.481
32	0.311	0.196	0.317	0.283	0.210	0.564	0.168	0.136	0.207	0.199	0.151	0.375	0.227	0.264	0.277	0.569	0.464	0.489
33	0.517	0.444	0.331	0.311	0.342	0.264	0.469	0.381	0.475	0.594	0.536	0.314	0.406	0.289	0.681	0.408	0.305	0.283
34	0.419	0.227	0.230	0.227	0.208	0.567	0.196	0.185	0.227	0.219	0.272	0.864	0.531	0.261	0.353	0.300	0.286	0.369
35	0.249	0.542	0.456	0.567	0.558	0.230	0.606	0.575	0.428	0.494	0.442	0.269	0.258	0.369	0.319	0.275	0.239	0.196
36	0.225	0.356	0.361	0.308	0.322	0.272	0.347	0.347	0.444	0.433	0.394	0.319	0.500	0.283	0.464	0.347	0.258	0.250

	19	20	21	22	23	24	25	26	27	28	29	30	31	32	33	34	35	36
1	0.165	0.272	0.325	0.397	0.325	0.221	0.286	0.250	0.289	0.327	0.289	0.380	0.536	0.311	0.517	0.419	0.249	0.225
2	0.350	0.162	0.162	0.216	0.322	0.406	0.204	0.303	0.542	0.208	0.311	0.280	0.333	0.196	0.444	0.227	0.542	0.356
3	0.369	0.381	0.358	0.414	0.314	0.367	0.202	0.222	0.517	0.221	0.258	0.210	0.389	0.317	0.331	0.230	0.456	0.361
4	0.272	0.303	0.316	0.406	0.283	0.361	0.213	0.244	0.572	0.199	0.261	0.207	0.389	0.283	0.311	0.227	0.567	0.308
5	0.199	0.241	0.252	0.258	0.317	0.250	0.213	0.241	0.461	0.225	0.436	0.244	0.297	0.210	0.342	0.208	0.558	0.322
6	0.202	0.561	0.575	0.608	0.369	0.224	0.381	0.428	0.185	0.366	0.238	0.151	0.347	0.564	0.264	0.567	0.230	0.272
7	0.336	0.142	0.170	0.204	0.261	0.392	0.170	0.283	0.642	0.191	0.241	0.378	0.353	0.168	0.469	0.196	0.606	0.347
8	0.297	0.107	0.153	0.153	0.250	0.344	0.133	0.227	0.575	0.185	0.367	0.347	0.342	0.136	0.381	0.185	0.575	0.347
9	0.467	0.216	0.249	0.333	0.331	0.556	0.182	0.328	0.367	0.168	0.264	0.364	0.461	0.207	0.475	0.227	0.428	0.444
10	0.408	0.185	0.227	0.269	0.400	0.444	0.193	0.330	0.436	0.167	0.319	0.558	0.483	0.199	0.594	0.219	0.494	0.433
11	0.389	0.151	0.142	0.173	0.331	0.450	0.170	0.381	0.431	0.207	0.286	0.450	0.417	0.151	0.536	0.272	0.442	0.394
12	0.286	0.375	0.333	0.392	0.367	0.327	0.531	0.353	0.225	0.575	0.417	0.224	0.403	0.375	0.314	0.864	0.269	0.319
13	0.684	0.405	0.303	0.328	0.386	0.628	0.422	0.275	0.269	0.453	0.358	0.286	0.411	0.227	0.406	0.531	0.258	0.500
14	0.182	0.181	0.269	0.289	0.286	0.239	0.174	0.176	0.353	0.199	0.252	0.224	0.339	0.264	0.289	0.261	0.369	0.283
15	0.378	0.272	0.291	0.305	0.558	0.453	0.205	0.367	0.336	0.272	0.367	0.311	0.517	0.277	0.681	0.353	0.319	0.464
16	0.300	0.600	0.742	0.753	0.469	0.247	0.303	0.258	0.199	0.258	0.297	0.244	0.531	0.569	0.408	0.300	0.275	0.347
17	0.255	0.500	0.569	0.633	0.350	0.227	0.275	0.241	0.188	0.236	0.308	0.247	0.394	0.464	0.305	0.286	0.239	0.258
18	0.289	0.525	0.653	0.661	0.383	0.252	0.325	0.252	0.176	0.255	0.303	0.207	0.481	0.489	0.283	0.369	0.196	0.250
19	1.000	0.506	0.378	0.361	0.353	0.717	0.341	0.366	0.266	0.367	0.328	0.294	0.381	0.213	0.353	0.289	0.247	0.639
20	0.506	1.000	0.700	0.700	0.450	0.328	0.406	0.319	0.188	0.456	0.356	0.151	0.311	0.533	0.225	0.369	0.179	0.394
21	0.378	0.700	1.000	0.807	0.444	0.275	0.333	0.269	0.202	0.277	0.300	0.297	0.458	0.528	0.339	0.322	0.266	0.339
22	0.361	0.700	0.807	1.000	0.411	0.336	0.339	0.306	0.244	0.278	0.305	0.269	0.414	0.572	0.336	0.375	0.247	0.394
23	0.353	0.450	0.444	0.411	1.000	0.392	0.364	0.461	0.286	0.342	0.478	0.297	0.486	0.464	0.625	0.364	0.325	0.467
24	0.717	0.328	0.275	0.336	0.392	1.000	0.294	0.347	0.422	0.300	0.336	0.352	0.461	0.222	0.492	0.297	0.291	0.745
25	0.341	0.406	0.333	0.339	0.364	0.294	1.000	0.241	0.176	0.531	0.406	0.170	0.342	0.322	0.241	0.567	0.196	0.311
26	0.366	0.319	0.269	0.306	0.461	0.347	0.241	1.000	0.367	0.236	0.209	0.300	0.392	0.597	0.428	0.325	0.322	0.325
27	0.266	0.188	0.202	0.244	0.286	0.422	0.176	0.367	1.000	0.188	0.277	0.342	0.325	0.181	0.369	0.207	0.664	0.333
28	0.367	0.456	0.277	0.278	0.342	0.300	0.531	0.236	0.188	1.000	0.280	0.176	0.358	0.224	0.216	0.656	0.193	0.292
29	0.328	0.356	0.300	0.305	0.478	0.336	0.406	0.209	0.277	0.280	1.000	0.389	0.381	0.272	0.533	0.375	0.255	0.467
30	0.294	0.151	0.297	0.269	0.297	0.352	0.170	0.300	0.342	0.176	0.389	1.000	0.389	0.145	0.478	0.188	0.439	0.353
31	0.381	0.311	0.458	0.414	0.486	0.461	0.342	0.392	0.325	0.358	0.381	0.389	1.000	0.280	0.622	0.425	0.333	0.447
32	0.213	0.533	0.528	0.572	0.464	0.222	0.322	0.597	0.181	0.224	0.272	0.145	0.280	1.000	0.316	0.311	0.233	0.219
33	0.353	0.225	0.339	0.336	0.625	0.492	0.241	0.428	0.369	0.216	0.533	0.478	0.622	0.316	1.000	0.344	0.361	0.561
34	0.289	0.369	0.322	0.375	0.364	0.297	0.567	0.325	0.207	0.656	0.375	0.188	0.425	0.311	0.344	1.000	0.213	0.277
35	0.247	0.179	0.266	0.247	0.325	0.291	0.196	0.322	0.664	0.193	0.255	0.439	0.333	0.233	0.361	0.213	1.000	0.339
36	0.639	0.394	0.339	0.394	0.467	0.745	0.311	0.325	0.333	0.292	0.467	0.353	0.447	0.219	0.561	0.277	0.339	1.000

2.2. *Using Actual Complete Experimental Data to Establish the Benchmark Classification System*

Fuzzy clustering analysis is used in this section to build up the benchmark classification system from the complete experimental data. Such a method can generate different outcomes in terms of the hierarchy and number of clusterings. The user, therefore, can choose the proper clustering as the evaluation basis to their knowledge of the similarity of the workpieces. The benchmark classification system features the flexible characteristics. This will be described in two parts: fuzzy clustering analysis and using aggregated comparison crisp numbers for workpiece clustering.

2.2.1. *Fuzzy clustering analysis*

To use the fuzzy clustering method, first, it is necessary to determine the similarity levels between the objects to be clustered. This similarity level in a fuzzy set can be thought of as a fuzzy relation, $\tilde{R}$. The fuzzy relation $\tilde{R}$ is both reflexive and symmetric. If it is also transitive, then $\tilde{R}$ is said to be a similarity relation,[19] denoted by $\tilde{\tilde{R}}$. By properly selecting an α-cut of the membership matrix ($\tilde{\tilde{R}}^{\alpha}$) for this similarity relation, one obtains an ordinary similarity relation. The equivalence of similarity relations among clustered objects can be utilized to carry out clustering. The definition below is a description of the fuzzy relation and similarity relation defined in fuzzy mathematics.

Definition 1. Suppose $X = \{x_1, x_2, \ldots, x_n\}$ is a referential set. A binary fuzzy relation $\tilde{R}$ on X is a fuzzy subset of the Cartesian product $X \times X$.

Let $\mu_{\tilde{R}} : X \times X \to [0, 1]$ denote the membership function of $\tilde{R}$ and let

$$\tilde{R} = [r_{ij}]_{nm}$$
$$r_{ij} = \mu_{\tilde{R}}(x_i, x_j) \quad x_i, x_j \in X.$$

A fuzzy relation $\tilde{R}$ on X is said to be reflexive if $\mu_{\tilde{R}}(x_i, x_j) = 1$ for all $x_i \in X$, $\tilde{R}$, is symmetric if $\mu_{\tilde{R}}(x_i, x_j) = \mu_{\tilde{R}}(x_j, x_i)$ for all $x_i, x_j \in X$, $\tilde{R}$ is (max-min) transitive if for any x_i, $x_k \in X$

$$\mu_{\tilde{R}}(x_i, x_k) \geq \max\{\min\{\mu_{\tilde{R}}(x_i, x_j), \mu_{\tilde{R}}(x_j, x_k)\}\} \qquad \forall x_j \in X.$$

If the fuzzy relation $\tilde{R}$ on X is reflexive, symmetric, and transitive, then $\tilde{R}$ is said to be a similarity relation on X, denoted by $\tilde{\tilde{R}}$.[19]

Because $\tilde{\tilde{R}}$ is a similarity relation on X that exists and is unique, then for any α-cut of membership matrix $\tilde{\tilde{R}}^{\alpha} = [r_{ij}^{\alpha}]$, in which $\alpha \in [0, 1]$

$$r_{ij}^{\alpha} = \begin{cases} 1 & r_{ij} \geq \alpha \\ 0 & r_{ij} < \alpha \end{cases} \tag{4}$$

are all similarity relations on X. These similarity relations can be used to classify the elements in X.[15,19] This kind of classification is called fuzzy clustering analysis.[25] Example 1 illustrates this type of analysis.

Example 1. If $\tilde{R}$ is a fuzzy relation on the set $X = \{x_1, x_2, \ldots, x_5\}$ with its membership matrix as follows, the elements in this matrix represent the level of similarity between x_i and x_j using fuzzy clustering analysis for classification.

$$\tilde{R} = \begin{array}{c} \\ x_1 \\ x_2 \\ x_3 \\ x_4 \\ x_5 \end{array} \begin{array}{ccccc} x_1 & x_2 & x_3 & x_4 & x_5 \\ \left(\begin{array}{ccccc} 1 & 0.1 & 0.8 & 0.5 & 0.3 \\ 0.1 & 1 & 0.1 & 0.2 & 0.4 \\ 0.8 & 0.1 & 1 & 0.5 & 0.3 \\ 0.5 & 0.2 & 0.5 & 1 & 0.3 \\ 0.3 & 0.4 & 0.3 & 0.3 & 1 \end{array} \right) \end{array}.$$

From the above definition, $\tilde{R}$ is reflexive and symmetric. We now need to calculate the max-min transitivity. The method of calculation is to search $\tilde{R}^2 = \tilde{R} \circ \tilde{R}, \ldots$, until $\tilde{R}^{2k} = \tilde{R}^k$ is found. In this example, we calculate $\tilde{R}^2$ and $\tilde{R}^4$ until $\tilde{R}^4 = \tilde{R}^2$ is found, and then $\tilde{R}^2$ is a similarity relation on X, denoted by $\hat{\tilde{R}}$.

$$\hat{\tilde{R}} = \tilde{R}^2 = \begin{array}{c} \\ x_1 \\ x_2 \\ x_3 \\ x_4 \\ x_5 \end{array} \begin{array}{ccccc} x_1 & x_2 & x_3 & x_4 & x_5 \\ \left(\begin{array}{ccccc} 1 & 0.3 & 0.8 & 0.5 & 0.3 \\ 0.3 & 1 & 0.3 & 0.3 & 0.4 \\ 0.8 & 0.3 & 1 & 0.5 & 0.3 \\ 0.5 & 0.3 & 0.5 & 1 & 0.3 \\ 0.5 & 0.4 & 0.3 & 0.3 & 1 \end{array} \right) \end{array}.$$

By using $\hat{\tilde{R}}$ for any α-cut ($\hat{\tilde{R}}^\alpha$), we can cluster all similarity relations. The clustering results for each of these similarity relations are shown in Fig. 7.

From the classification mentioned above, it is clear that different values of "α" means that the number of clustering and the elements in each clustering will vary. When the value is high, only elements of X that have a high degree of similarity will be placed near to each other; when the value is low, elements that have a low degree

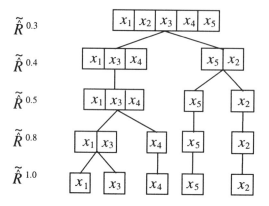

Fig. 7.   Using fuzzy clustering for Example 1.

of similarity will also be placed together. The value of "α" can be used to represent subjects' judgements of the degree of similarity between various workpieces. Different values of "α" will produce different numbers of clusters containing different elements. Thus, the value of "α" can be seen as an index of flexibility. Users can use the value of "α" to adjust the number of groups in the resulting classification according to their practical needs.

2.2.2. *Using aggregated comparison crisp numbers for workpiece clustering*

After the subjects compared the 36 sample workpieces, the crisp numbers obtained from the fuzzy relations representing the comparisons between the workpieces were determined through membership function aggregation and defuzzification. These crisp numbers (Table 2) can be seen as the fuzzy relation on $\tilde{R}$ of the 36 workpieces in the Cartesian product $X \times X$ of a given set $X = \{x_1, x_2, \ldots, x_{36}\}$. As in the example above, this fuzzy relation is reflexive and symmetric. If the calculation satisfies the max-min transitivity and we find $\tilde{R}^{16} = \tilde{R}$, then this fuzzy relation is also a similarity relation on X. Arranging the α-cuts on $\tilde{R}$ in order from largest to smallest $(1, 0.864, 0.823, 0.818, \ldots, 0.533)$, we obtain different similarity relations on X. The classification results corresponding to different similarity relations for the 36 sample workpieces, which are defined as the benchmark classification system and denoted as C, are shown in Fig. 8. This chart represents the entire benchmark classification system, and the different classifications shown, which are generated using different similarity relations, can each be used as a benchmark classification. The user can select a particular benchmark classification or number of workpiece groups (and the corresponding α-level) depending on his/her intuitive judgement of the degree of similarity between different workpieces. The indices then allow us to measure the level of consistency between the benchmark classification selected and the test classification.

3. The Evaluation of the Benchmark Classification System

The purpose of the benchmark classification system is to evaluate the utility of the automated workpiece classification system. In this section, we developed two indices — the appropriate number of workpiece groups and the degree of consistency between the two classifications — with which to evaluate the utility of test classifications. We will explain these indices and describe how we applied them to evaluate the automated workpiece classification technique developed by Wu and Jen.[23] Features of the benchmark classification system and the difficulties in practical application are also discussed.

3.1. *Index of appropriate number of workpiece groups*

This index is used to compare the number of groups classified by the test classification system with the number in the benchmark cluster classification, and thus to

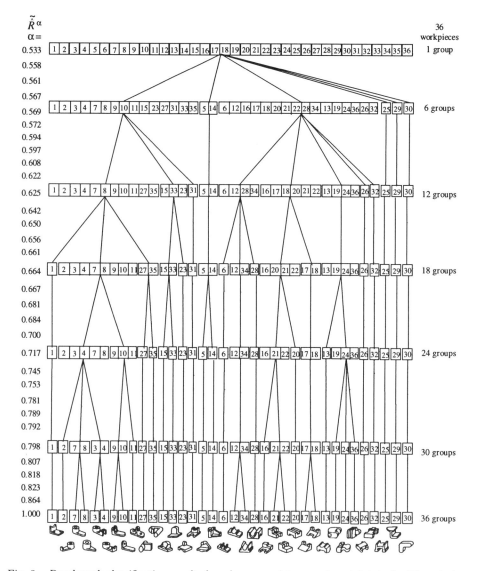

Fig. 8.   Benchmark classification results based on complete experimental data for 36 workpieces and denoted as C.

determine whether the test classification generates too many or too few groups. If too many groups are formed, the criteria for similarity applied in the system is too strict; if too few are formed, the criteria is too loose. The formula for calculating the index is shown below:

$$r = \begin{cases} \dfrac{q - 1}{q^* - 1} & (1 \leq q \leq q^*) \\ \dfrac{x - q}{x - q^*} & (q^* \leq q \leq x) \end{cases} \tag{5}$$

where

q is the number of groups in the test classification system

q^* is the user-specified number of groups in the benchmark classification

x is the number of workpieces being classified

r is the index of appropriate number of workpiece groups.

The level of consistency between the number of groups in the test classification and in the benchmark classification as measured by this formula is indicated by a value from 0 to 1, with 0 representing the largest difference and 1 representing completely identical groups. The following example illustrates the application of this index.

Example 2. Consider a sample containing 36 workpieces ($x = 36$). If the user selects a benchmark classification with 30 groups and there are 24 groups in the test classification, then the index of the appropriate number of workpiece groups for the test classification will be:

$$r = \frac{24 - 1}{30 - 1} = 0.79.$$

By the same principle, if the number of groups in the benchmark classification is 24, then the index for the test classification will be 1.00. If the number of groups in the benchmark classification is 18, then the index for the test system will be 0.67. If the number of groups in the benchmark classification is 12, the index will be 0.55. If the number of groups in the benchmark classification is 6, the index will be 0.40.

3.2. *The degree of consistency between two classifications*

At present, no useful method is available for comparing the degree of consistency between two different classification systems. Therefore, the aim of this section is to develop a heuristic algorithm for measuring the consistency between the classification results for the same groups in the test classification and benchmark classification. The idea is to calculate the ratio of the sum of the number of workpieces in each corresponding group in the two classifications that are the same over the total number of workpieces within each group. This ratio will be referred to as the index of consistency, the average degree of similarity within each group in the test classification and the corresponding group in the benchmark classification. As described earlier, a one-to-one correspondence is found between groups in the two classifications, both in the same and different number of groups, by starting from the classification with fewer groups. Each pair of corresponding groups from the two classifications is called a corresponding pair. The algorithm used to calculate this index is explained below.

Step 1. Arrange the classifications being compared in a matrix.

(1) Suppose there are p workpieces, $x_1, x_2, \ldots, x_p$, classified by means of the two classification methods. The first method produces m groups. The classification

Table 3.   Results of two classification arranged in a matrix.

2nd Classification 1st Classification	B_1	$\cdots$	B_j	$\cdots$	B_n	
A_1	c_{11}	$\cdots$	c_{1j}	$\cdots$	c_{1n}	d_1
$\vdots$	$\vdots$		$\vdots$		$\vdots$	$\vdots$
A_i	c_{i1}	$\cdots$	c_{ij}	$\cdots$	c_{in}	d_i
$\vdots$	$\vdots$		$\vdots$		$\vdots$	$\vdots$
A_m	$cm1$	$\cdots$	c_{mj}	$\cdots$	c_{mn}	d_m

results are denoted by $A_1, \ldots, A_i, \ldots, A_m$; A_i denotes the set of workpieces within group i. The second classification method produces n groups. The classification results are denoted by $B_1, \ldots, B_j, \ldots, B_n$; B_j denotes the set of workpieces within group j. Suppose $m \leq n$.

(2) The classification results from the first method are arranged in columns and those from the second method in rows (see Table 3).

Step 2. Let c_{ij} in the 2D matrix be the number of workpieces included in both A_i and B_j, that is, $c_{ij} = N(A_i \cap B_j)$.

Step 3. Determine the correspondences between groups.

(1) Arrange all the values of c_{ij} in rank from large to small, $i = 1, 2, \ldots, m$; $j = 1, 2, \ldots, n$.

(2) Take the largest value of c_{ij}. The corresponding A_i and B_j are the first corresponding pair of groups from the two classification methods. Delete the row i and column j that intersect at this c_{ij} in the matrix.

(3) If there are two or more identical largest values of c_{ij}, then compare the second largest value in the ith row and the jth column corresponding to the largest c_{ij}, and choose the smallest of these values. Then delete the row and column corresponding to the value chosen. If the second largest values are also identical, then compare the third largest values, and so on until a usable value is obtained. If the last values for comparison are still identical, then choose each value and proceed the calculation.

(4) Repeat the above procedure for all undeleted c_{ij} until all the rows and columns in the matrix have been deleted.

(5) During the calculation, if there are no identical values, the corresponding groups chosen in the step will be the corresponding groups of these two classifications.

(6) If there exist identical values in the calculation process, we need to exhaustively list the possible corresponding groups for these two classifications. The one that has the largest number of workpieces for each classification will be chosen as the corresponding group for these two classifications.

Step 4. Calculate the degree of consistency between two classifications.

(1) Let d_i denote the value of c_{ij} for each pair of corresponding groups.
(2) Let h denote the degree of consistency between two classifications. That is the number of workpieces shared by all corresponding groups over the total number of workpieces. The value of h is calculated as follows:

$$h = \frac{\sum_{i=1}^m d_i}{p}. \tag{6}$$

The index of the degree of consistency between two classifications falls between $1/p$ and 1, that is, $1/p \leq h \leq 1$. If there is a large number of workpieces in a group; that is, $p \gg 0$, then the value will be between 0 and 1. The following example shows how the index works.

Example 3. Consider a set of 10 sample workpieces $(x_1, x_2, \ldots, x_{10})$. Suppose there are 4 groups in the first classification and 5 in the second, as shown in Table 4. We will use the above algorithm to calculate the degree of consistency between the two sets of classification results.

According to Step 1 and Step 2 of the algorithm, we calculate the number of workpieces c_{ij} that appear in any pair of groups A_i and B_j, as shown in Table 4. In this case, the value of c_{11} is 1, which means there is only one workpiece x_1, that is included in both A_1 and B_1. In Step 3, we arrange all values of c_{ij} in order from largest to smallest. The largest value is 2, and there are two sets, c_{22}, and c_{23} with this value. So we then compare the second-largest value in each of the rows and columns corresponding to these two sets, and we find that in each case the second-largest value is 2. Hence we must go on to compare the third-largest values which are still identical and equal to 1. We hereby need to list all possible corresponding groups. First, we choose c_{22}, from which A_2 and B_2 is the first set of would-be-corresponding groups. Delete the second row and the second column, we get five largest values each of which is equal to 1. Comparing the second largest value, we get A_1 and B_1 as the second would-be-corresponding group. Repeating the procedure until all c_{ij}'s are deleted from the matrix, we now find three would-be-corresponding groups, A_1 and B_1, A_2 and B_2, and A_4 and B_3. Among these

Table 4. The two sets of classification results in Example 3.

2nd Classification 1st Classification	$B_1 =$ $\{x_1, x_9\}$	$B_2 =$ $\{x_2, x_3, x_4\}$	$B_3 =$ $\{x_5, x_6, x_7,\}$	$B_4 =$ $\{x_8\}$	$B_5 =$ $\{x_{10}\}$
$A_1 = \{x_1\}$	1	0	0	0	0
$A_2 = \{x_2, x_4, x_5, x_6\}$	0	2	2	0	0
$A_3 = \{x_3\}$	0	1	0	0	0
$A_4 = \{x_7, x_8, x_9, x_{10}\}$	1	0	1	1	1

Table 5.   The corresponding groups to the largest value c_{22} for Example 3.

2nd Classification / 1st Classification	$B_1 =$ $\{x_1, x_9\}$	$B_2 =$ $\{x_2, x_3, x_4\}$	$B_3 =$ $\{x_5, x_6, x_7\}$	$B_4 =$ $\{x_8\}$	$B_5 =$ $\{x_{10}\}$	Number of Work-pieces in Corresponding Groups
$A_1 = \{x_1\}$	①	0	0	0	0	1
$A_2 = \{x_2, x_4, x_5, x_6\}$	0	②	2	0	0	2
$A_3 = \{x_3\}$	0	1	0	0	0	
$A_4 = \{x_7, x_8, x_9, x_{10}\}$	1	0	①	1	1	1

Table 6.   The corresponding groups to the largest value c_{23} for Example 3.

2nd Classification / 1st Classification	$B_1 =$ $\{x_1, x_9\}$	$B_2 =$ $\{x_2, x_3, x_4\}$	$B_3 =$ $\{x_5, x_6, x_7\}$	$B_4 =$ $\{x_8\}$	$B_5 =$ $\{x_{10}\}$	Number of Work-pieces in Corresponding Groups
$A_1 = \{x_1\}$	①	0	0	0	0	1
$A_2 = \{x_2, x_4, x_5, x_6\}$	0	2	②	0	0	2
$A_3 = \{x_3\}$	0	①	0	0	0	
$A_4 = \{x_7, x_8, x_9, x_{10}\}$	1	0	1	①	1	1

sets, the number of workpieces that can be found in the intersection part of the corresponding sets is 4, as shown in Table 5.

Moreover, c_{23} is selected from the largest values. Four corresponding groups A_1 and B_1, A_2 and B_3, A_3 and B_2, and A_4 and B_4 can be found from the same calculation procedure. The number of workpieces in the intersection of corresponding groups is 5, as shown in Table 6.

The corresponding groups can be obtained either from Table 5 or from Table 6. Comparing the results, four corresponding groups: A_1 and B_1, A_2 and B_3, A_3 and B_2, and A_4 and B_4 are identified from Table 6. Finally, in Step 4, the number of workpieces shared between each of these pairs of corresponding groups is 1, 2, 1, and 1, respectively. Thus $h = (1+2+1+1)/10 = 0.5$, which is the degree of consistency between these two classification schemes.

3.3. A practical example

To demonstrate the use of the proposed benchmark classification system, we took Wu and Jen's[23] automated classification technique as the test classification system. It divides the 36 sample workpieces into 24 groups, as shown in Table 7. We then used the benchmark classification system to evaluate this system.

If the user applies a strict criterion of similarity in classifying the workpieces, then a classification obtained from a high value for "α" should be used as the benchmark to evaluate the test classification (see Fig. 8). If a looser criterion is needed, then a classification obtained with a lower value for "α" can be chosen. In this research, we chose $\alpha = 0.798, 0.717, 0.664, 0.625$ and 0.599 to represent five

Table 7. Comparison of results of benchmark classification and automatic classification.

Classification Type		Workpiece Clustering	Number of Groups	Appropriate Number of Workpieces Groups (r)	Degree of Consistency Between Two Classification (h)
Benchmark classification (denotes the user's judgement of workpiece similarity)	α = 0.798	1 2 [7 8] [3 4] 5 6 [9 10] 11 [12 34] 13 14 15 16 [17 18] 19 20 [21 22] 23 24 25 26 27 28 29 30 31 32 33 35 36	30	0.79	0.72
	0.717	1 | [2 3 4 7 8] [5 6] [9 10 11] [12 34] [13 14 15] [16 21 22] 20 [17 18] [19 24 36] [23 25 26 27 28 29 30 31 32 33 35]	24	1.00	0.64
	0.664	1 [2 3 4 7 8 9 10 11] 6 [12 34] [5 14] [13 19 24 36] [15 33] [16 20 21 22] [17 18] [23 25 26] 27 [28 29 30 31 32] 35	18	0.67	0.53
	0.625	[1 2 3 4 7 8 9 10 11 27 35] [5 14] [6 12 28 34] [13 19 24 36] [15 33 23] [16 17 18 20 21 22] [25 26 29 30 31 32]	12	0.55	0.42
	0.599	[1 2 3 4 7 8 9 10 11 15 23 27 31 33 35] [5 14] [6 12 16 17 18 20 21 22 28 34] [13 19 24 36 26 32] [25 29 30]	6	0.40	0.25
Automatic classification system	Wu and Jen (1996)	[2 4] 3 [7 8 10 30] [12 13 34] [1 9 17 22] [6 33] [11 27] [5 14 15 16] [18 21] [19 20] [23 29] [24 25 26 28 31 32 35 36]	24		

Remarks
1. ▢ denotes groups of workpieces
2. Workpieces not included within ▢ each constitute their own group

different levels of strictness in judging similarity and used the two indices described in previous sections to evaluate the utility of the test classification. The results of our calculations are shown in Table 7. When $\alpha = 0.599$, 0.625, and 0.664, there are 6, 12, 18 groups in the benchmark classification. In these three cases, there are far fewer groups than the 24 identified by the test classification. The index of the appropriate number of groups is only 0.40, 0.55 and 0.67, respectively, for these three cases, which means that the test classification applies a stricter criterion of similarity within groups than the benchmark classification does. The index of degree of consistency between two classifications is 0.25, 0.42, and 0.53 respectively, which indicates that the similarity criteria applied by the benchmark classification are very different to those applied by the test classification. When $\alpha = 0.717$ on the other hand, there are 24 groups in the benchmark classification, which is the same as the number in the test classification. Thus, in this case the criteria of similarity applied by the benchmark classification and the test classification system in clustering the workpieces are equally strict. However, the index of degree of consistency between two classifications is only 0.64, indicating that the specific criteria applied by the two systems are very different. This may be because the test classification employs three 2D orthogonal drawings for workpiece clustering first and then aggregate the results. Using aggregated information in this way is more difficult than directly comparing global shape information and then clustering the workpieces. When "α" = 0.798, the benchmark system produces 30 groups, and the index of the appropriate number of groups is 0.79, which means that the test classification is now applying a looser criterion of similarity than the benchmark classification. The index of degree of consistency between two classifications is 0.72. This value is higher than the other values of "α" because the benchmark system now divides the workpieces into 30 groups. Since many of these groups now include only one workpiece, the index of consistency within each workpiece group will tend to be higher.

3.4. Difficulties in applying the benchmark classification to practical cases

The effectiveness of the automated classification system can be evaluated using the benchmark classification system on the basis of the criteria developed in this study. The most effective automated classification system can be selected using this method. The workpiece clustering database obtained from the classification procedure is more compatible to the user's mental model. It is, therefore, helpful in the information retrieval and application for the user of various purposes.

It is, however, a time-consuming task for the subject to perform the workpiece similarity pair-comparison task. To make the matter worse, the subjects' judgement might be deteriorated if the number of comparison tasks is too large or the comparison task lasts for a long time. To cope with these difficulties, one way is to carefully bring the experiment under control. Another aggressive way is to develop a systematic method to reduce the number of pair comparisons but still maintain the proper data reliability. In other words, a small number of typical workpieces are selected from among the sample workpieces, and each of these is compared with each sample workpiece to obtain the partial experimental data. Then the complete experimental data can be estimated with a mathematical inference method. The classification result obtained in such a procedure is defined as the lean classification. As long as it meets the utility requirement, the lean classification can substitute for the benchmark classification.

4. Using Partial Experimental Data to Establish the Lean Classification System

In this section, partial experimental data from the workpiece similarity pair comparison is used to build up the lean classification system. The feasibility of the lean classification system depends on its utility. This chapter is made up of three parts as shown in Fig. 9. Section 4.1 deals with the process in which partial experimental data are selected from the complete experimental data. In Sec. 4.2, on the contrary, the partial experimental data are used to infer the complete experimental data. Finally, Sec. 4.3 describes how to build up the lean classification system by using the quasi-complete experimental data.

4.1. *Selecting partial experimental data from the complete experimental data*

In a practical environment, if there are an excessive number of sample workpieces, it will be time-consuming and costly to carry out complete experimental

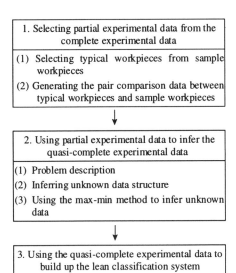

Fig. 9.   The procedure to build up the lean classification system.

comparisons of the workpieces. Consider a set of 1000 sample workpieces. The complete experimental testing would involve about 0.5 million pair comparisons. If each of these comparisons takes 10 seconds and the subjects worked 8 hours a day, it would take 173 working days to complete the comparison. In place of this type of exhaustive testing, in the proposed method a small number of typical workpieces with different shapes are selected from among the sample workpieces. Each of these typical workpieces is then compared with each of the sample workpieces to obtain partial comparison data. Then these partial data are extrapolated to estimate the results of the complete experimental data for all of the samples. These data are called the quasi-complete experimental data. The classification results obtained using the estimated data are called the lean classification. The degree of consistency between the lean classification and the benchmark classification will indicate the effectiveness of the method used to produce the lean classification.

4.1.1. Selecting typical workpieces from sample workpieces

Typical workpieces are selected from the sample workpieces using a binary clustering method. First, the sample workpieces are classified by selecting a user at random and having the user hierarchically divide the workpieces into subgroups on the basis of the similarity of their global shape (see Fig. 10). Second, groups in the hierarchy are selected at random according to the number of typical workpieces needed: when a lean classification is to be established, the exact number of groups sampled from the relative hierarchy (h) will depend on the number of typical workpieces (m) required. The relationship between m and h can be expressed by $2^{h-1} < m \leq 2^h$.

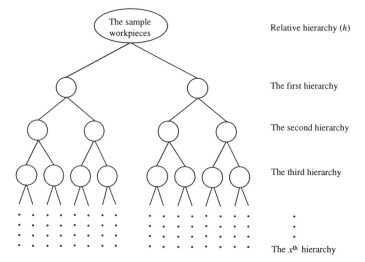

Fig. 10.   Selecting typical workpieces from sample workpieces using binary clustering.

Then one workpiece is chosen at random from each selected group, and the resulting workpieces are used as the typical workpieces. For example, if a lean classification is to be established on the basis of six typical workpieces, then the typical workpieces should be randomly selected from any six of the eight groups in the third level of the hierarchy (see Fig. 10). Selecting typical workpieces from different groups in this way helps to ensure that there are significant differences between the typical workpieces.

4.1.2. *Generating the pair comparison data between typical workpieces and sample workpieces*

One subject was randomly selected from an aircraft plant to classify the 36 workpieces mentioned earlier according to the similarity of their global shape. The classification task was proceeded by a binary hierarchical approach in which the sixth hierarchy was reached and there is only one workpiece in the hierarchy level. The record indicated that the workpieces 24th and 14th were selected from the first hierarchy by the subject. From the second hierarchy, workpieces 7th, 4th, 5th, and 31st were chosen. Similarly, workpieces 11th, 24th, 26th, 3rd, 14th, 1st, 18th, and 12th were picked from the third hierarchy. In other words, if we need to select two typical workpieces for the partial experimental data, the similarity relations of the workpieces 24th and 14th rows will be chosen from the complete experimental data in Table 2. In the same way, workpieces in the 11th, 24th, 26th, 3rd, 14th, 1st, 18th, and 12th rows will be selected if we need eight typical workpieces for the partial experimental data.

4.2. *Using partial experimental data to infer the quasi-complete experimental data*

This section describes a method for using partial experimental data to infer the quasi-complete experimental data. The aim is to utilize the known data obtained from pair comparisons between each typical workpiece and all of the sample workpieces to infer unknown data concerning the degree of similarity between all of the non-typical workpieces. The combination of the known and inferred data will then be equivalent to using the quasi-complete experimental data. The procedure for inferring the unknown data structure is described in the following three parts: (1) problem description, (2) inferring unknown data structure, and (3) using the max-min method to infer the unknown data.

4.2.1. *Problem description*

For convenience, the levels of similarity between all workpieces will be expressed in the format of a matrix S, in which $x_1, x_2, \ldots, x_n$ are sample workpieces;

$x_1, x_2, \ldots, x_m$ are typical workpieces; and $x_{m+1}, x_{m+2}, \ldots, x_n$ are non-typical work-pieces.

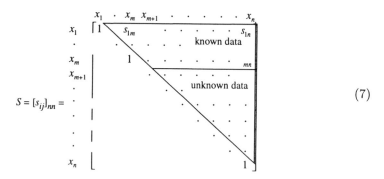

$$S = [s_{ij}]_{nn} = \qquad\qquad\qquad\qquad\qquad\qquad\qquad\qquad\qquad (7)$$

In formula (7), an element s_{ij} of matrix S represents the degree of similarity between x_i and x_j. Because $s_{ij} = s_{ji}$ and $s_{ii} = 1$, the matrix is symmetric. Therefore, only the upper part of the data need be considered. In s_{ij}, if $i = 1, 2, \ldots, m$ and $j = i + 1, \ldots, n$, then the elements in the trapezoid part are known data and the elements in the remaining triangular part are unknown data.

4.2.2. *Inferring the unknown data structure*

The basic idea in this research is to infer unknown data about similarity levels between non-typical workpieces using the degree of similarity between two workpieces and a third workpiece to infer the degree of similarity between the two workpieces, as shown in Fig. 11(a). Consider a typical workpiece x_k. According to known data, the degree of similarity between x_k and a non-typical workpiece x_p is s_{pk} and that between x_k and a third workpiece x_q is s_{kq}. The inference method proposed here allows us to estimate s_{pq}, the degree of similarity between x_p and x_q. This degree of similarity will be denoted by $s_{pq(k)}$, which means that the degree of similarity between x_p and x_q was derived from x_k.

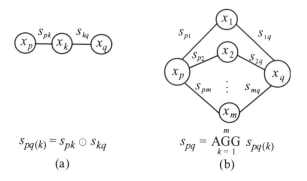

Fig. 11. s_{pq} is inferred from m similarity levels between x_p and x_q.

Because there are m typical workpieces, each unknown datum s_{pq} can be inferred from m typical workpieces; that is, $s_{pq(k)}$, $k = 1, 2, \ldots, m$. To utilize the existing data to infer the degree of similarity between x_p and x_q, mathematical methods must be used to aggregate these m data, as shown in Fig. 11(b). Accordingly, the above-mentioned inference process can be described as follows:

$$s_{pq} = \underset{k=1}{\overset{m}{\text{AGG}}} s_{pk} \odot s_{kq}$$

$$= \underset{k=1}{\overset{m}{\text{AGG}}} s_{pq(k)} \qquad (8)$$

where

$\odot$ represents the inference method

AGG represents the method of aggregation

s_{pk} represents the degree of similarity between x_p and x_k

s_{kq} represents the degree of similarity between x_k and x_q

$s_{pq(k)}$ represents the degree of similarity between x_p and x_q, which is inferred from kth typical workpieces

s_{pq} represents the degree of similarity between two non-typical workpieces x_p and x_q, which is inferred from m typical workpieces.

4.2.3. *Using max-min method to infer unknown data*

Fuzzy set theory is used to infer the unknown data. First, the global shape attributes of each typical workpiece are used as a template. The global shape similarity between any two non-typical workpieces and this template can be regarded as a fuzzy relationship between these non-typical workpieces and the template, with a value of [0, 1]. Second, the degree of similarity between the two non-typical workpieces can be inferred by performing a max-min composition of the fuzzy relationship between the two non-typical workpieces and all typical workpieces. Finally, this method can then be used to infer the degree of similarity between all non-typical workpieces.

The max-min composition of two fuzzy relations is defined as follows:[25]

Definition 2. Suppose $\tilde{R}$ is a fuzzy relation on the Cartesian product $X \times Y$ and $\tilde{S}$ is a fuzzy relation on $Y \times Z$. Then the max-min composition of $\tilde{R}$ and $\tilde{S}$, denoted by $\tilde{R} \circ \tilde{S}$, is a fuzzy relation on $X \times Z$ and the membership functions

$$\mu_{\tilde{R} \circ \tilde{S}}(x, z) = \max_{y \in Y} \min[\mu_{\tilde{R}}(x, y), \mu_{\tilde{S}}(y, z)] \qquad \forall x \in X, z \in Z.$$

The above definition can be used to infer unknown data on degrees of similarity between non-typical workpieces. We can let $K = \{x_1, x_2, \ldots, x_m\}$ be a referential set of typical workpieces and $P = Q = \{x_{m+1}, x_{m+2}, \ldots, x_n\}$ be a referential set of non-typical workpieces. Suppose $\tilde{S}_{PK} = [s_{pk}]_{(n-m)m}$ is a fuzzy relation on $P \times K$ in which $p = m + 1, \ldots, n$, $k = 1, 2, \ldots, m$, and $\tilde{S}_{KQ} = [s_{kq}]_{m(n-m)}$ is a fuzzy relation on $K \times Q$ in which $k = 1, 2, \ldots, m$, $q = m + 1, \ldots, n$. Then the composition of these two fuzzy relations can determine a fuzzy relationship between non-typical workpieces $P \times Q$. That is,

$$\tilde{S}_{PQ} = \tilde{S}_{PK} \circ \tilde{S}_{KQ}$$
$$= [s_{pk}]_{(n-m)m} \circ [s_{kq}]_{m(n-m)}$$
$$= [s_{pq}]_{(n-m)(n-m)} \tag{9}$$

where

$$s_{pq} = \max_{1 \leq k \leq m} \min(s_{pk}, s_{kq}). \tag{10}$$

In formula (9), $\circ$ represents the operation of max-min composition, and in formula (10), min is the inference method ($\odot$) and max, the method of aggregation (AGG). To infer the unknown data, the value of the diagonal in $\tilde{S}_{PQ}$ should be 1 to meet the reflexivity requirement. Therefore, when this fuzzy relationship is used to make inferences, its membership function should be partially revised as shown in formula (11):

$$s_{pq} = \begin{cases} \max_{1 \leq k \leq m} \min(s_{pk}, s_{kq}) & \text{when } m+1 \leq p \neq q \leq n \\ 1 & \text{when } m+1 \leq p = q \leq n. \end{cases} \tag{11}$$

As an example, consider five workpieces $(x_1, x_2, \ldots, x_5)$ and two typical workpieces (x_1, x_2), by means of which partial experimental data based on subjects' similarity comparisons will be used to infer the quasi-complete experimental data. Suppose the partial experimental data are denoted by matrix P.

$$P = \begin{array}{c} \\ x_1 \\ x_2 \end{array} \begin{array}{ccccc} x_1 & x_2 & x_3 & x_4 & x_5 \\ \left(\begin{array}{ccccc} 1 & 0.1 & 0.8 & 0.5 & 0.3 \\ & 1 & 0.1 & 0.2 & 0.4 \end{array} \right). \end{array} \tag{12}$$

Then this method infers the quasi-complete experimental data as shown in matrix R. Take s_{34} for example. From formulas (8) and (11), we know that $s_{34(1)} = \min(0.8, 0.5) = 0.5$ and $s_{34(2)} = \min(0.1, 0.2) = 0.1$. Therefore, $s_{34} = \max(s_{34(1)}, s_{34(2)}) = 0.5$ (as shown in Fig. 11).

$$R = \begin{array}{c} \\ x_1 \\ x_2 \\ x_3 \\ x_4 \\ x_5 \end{array} \begin{array}{ccccc} x_1 & x_2 & x_3 & x_4 & x_5 \\ \left(\begin{array}{ccccc} 1 & 0.1 & 0.8 & 0.5 & 0.3 \\ & 1 & 0.1 & 0.2 & 0.4 \\ & & 1 & 0.5 & 0.3 \\ & & & 1 & 0.3 \\ & & & & 1 \end{array} \right). \end{array} \tag{13}$$

4.3. Using the quasi-complete experimental data to build up the lean classification system

To build up the lean classification system, partial experimental data of different numbers of typical workpieces are selected from 36 sample workpieces through the max-min method. Similar to the complete experimental data from the total 36 sample workpieces, these data are defined as the quasi-complete experimental data, denoted as $\tilde{\underline{R}}_x$ ($\tilde{\underline{R}}_x$ represents the quasi-complete experimental data inferred by x typical workpieces). With the quasi-complete experimental data, the clustering result of these sample workpieces can be exhaustively listed through the fuzzy clustering method so as to generate the lean classification system for the different typical workpieces. Such a lean classification system is denoted as C_x (C_x means the lean classification system built up from x typical workpieces).

It is clear that different numbers of typical workpieces will result in different quasi-complete experimental data, and consequently build up their corresponding lean classification systems. Consider an example of partial experimental data based on comparisons with eight typical workpieces (the 11th, 24th, 26th, 3rd, 14th, 1st, 18th and 12th workpieces in Fig. 2) collected from 30 subjects. Through the global shape similarity comparison of the typical workpieces with the other sample workpieces, the level of similarity among the 36 workpieces can be inferred by using the max-min method. The data inferred by eight typical workpieces can be denoted as $\tilde{R}_8$ (see Table 8), meaning the quasi-complete experimental data inferred by eight typical workpieces. Taking the inferred quasi-complete experimental data to calculate the max-min transitivity, we find that $\tilde{R}_8^8 = \tilde{R}_8$, and then this fuzzy relation is just a similarity relation on X. Taking the α-cuts $(1, 0.864, 0.823, 0.781, \ldots, 0.417)$ in decreasing order of similarity, we find that different α-cuts of the membership matrix $(\tilde{\underline{R}}_8^\alpha)$ yield different numbers of groups in each hierarchy. An exhaustive list of the clustering results gives us the lean classification established by these eight typical workpieces, which is denoted as C_8 (see Fig. 12). In the same way, another quasi-complete experimental data can be inferred by ten typical workpieces and can be denoted as $\tilde{\underline{R}}_{10}$. Finally, a lean classification system can be obtained from the quasi-complete experimental data and can be denoted as C_{10}.

5. Measuring the Effectiveness and Efficiency of the Lean Classification System

After establishing the lean and benchmark classifications, we compare them to determine the degree of consistency between these two systems. We then evaluate the effectiveness and efficiency of the lean classification system, which is determined by its consistency with the benchmark classification system and the extent to which it reduces the cost of establishing a classification. These issues are discussed in the following subsections.

5.1. *Consistency of the lean classification system with the benchmark classification system*

The algorithm delineated in this section is the same as that in Sec. 3.2 except that it is only applied to the consistency for the case when the number of clustering is

Table 8. Quasi-complete experimental data inferred by eight typical workpieces, $\bar{R}_8$.

	1	2	3	4	5	6	7	8	9	10	11	12	13	14	15	16	17	18
1	1.000	0.403	0.356	0.467	0.250	0.464	0.428	0.300	0.489	0.650	0.558	0.425	0.236	0.294	0.275	0.389	0.339	0.369
2	0.428	1.000	0.742	0.742	0.408	0.403	0.675	0.667	0.428	0.428	0.428	0.403	0.406	0.383	0.406	0.389	0.339	0.369
3	0.356	0.742	1.000	0.823	0.408	0.286	0.675	0.667	0.397	0.411	0.403	0.250	0.252	0.361	0.328	0.336	0.302	0.241
4	0.467	0.742	0.823	1.000	0.417	0.464	0.675	0.667	0.467	0.467	0.467	0.425	0.361	0.417	0.408	0.389	0.339	0.369
5	0.356	0.408	0.408	0.417	1.000	0.294	0.408	0.408	0.397	0.408	0.403	0.294	0.286	0.667	0.408	0.336	0.302	0.294
6	0.464	0.403	0.356	0.464	0.294	1.000	0.428	0.322	0.464	0.464	0.464	0.642	0.569	0.294	0.433	0.433	0.433	0.433
7	0.533	0.675	0.675	0.675	0.408	0.428	1.000	0.667	0.533	0.533	0.533	0.425	0.392	0.380	0.392	0.389	0.339	0.369
8	0.442	0.667	0.667	0.667	0.408	0.322	0.667	1.000	0.442	0.442	0.442	0.327	0.344	0.389	0.389	0.336	0.302	0.300
9	0.558	0.428	0.403	0.467	0.397	0.464	0.533	0.442	1.000	0.736	0.736	0.425	0.556	0.361	0.453	0.389	0.339	0.369
10	0.650	0.428	0.411	0.467	0.408	0.464	0.533	0.442	0.736	1.000	0.781	0.425	0.444	0.361	0.444	0.389	0.339	0.369
11	0.558	0.428	0.403	0.428	0.230	0.322	0.533	0.442	0.736	0.781	1.000	0.319	0.286	0.255	0.325	0.156	0.095	0.147
12	0.425	0.241	0.250	0.250	0.286	0.642	0.239	0.191	0.330	0.291	0.319	1.000	0.569	0.199	0.411	0.342	0.330	0.339
13	0.425	0.406	0.367	0.361	0.286	0.569	0.392	0.344	0.556	0.444	0.436	0.569	1.000	0.255	0.453	0.342	0.330	0.339
14	0.294	0.383	0.361	0.417	0.667	0.185	0.380	0.389	0.264	0.255	0.255	0.199	0.190	1.000	0.408	0.247	0.257	0.280
15	0.411	0.406	0.367	0.408	0.408	0.433	0.392	0.389	0.453	0.444	0.436	0.411	0.453	0.408	1.000	0.444	0.444	0.444
16	0.389	0.389	0.356	0.389	0.336	0.433	0.389	0.336	0.389	0.389	0.389	0.389	0.342	0.336	0.444	1.000	0.639	0.639
17	0.369	0.339	0.339	0.339	0.302	0.433	0.339	0.302	0.339	0.339	0.339	0.339	0.330	0.302	0.444	0.639	1.000	0.864
18	0.369	0.145	0.241	0.303	0.294	0.433	0.142	0.145	0.221	0.184	0.147	0.339	0.249	0.280	0.444	0.639	0.864	1.000
19	0.389	0.406	0.389	0.389	0.369	0.366	0.392	0.389	0.556	0.444	0.436	0.353	0.628	0.361	0.453	0.336	0.302	0.289
20	0.375	0.381	0.381	0.381	0.381	0.433	0.381	0.381	0.381	0.381	0.381	0.375	0.375	0.361	0.444	0.525	0.525	0.525
21	0.369	0.358	0.358	0.358	0.358	0.433	0.358	0.358	0.358	0.358	0.358	0.339	0.333	0.358	0.444	0.639	0.653	0.653
22	0.397	0.414	0.414	0.414	0.408	0.433	0.414	0.414	0.397	0.411	0.403	0.397	0.392	0.361	0.444	0.639	0.661	0.661
23	0.369	0.392	0.367	0.361	0.314	0.428	0.392	0.344	0.392	0.392	0.392	0.367	0.392	0.314	0.392	0.383	0.383	0.383
24	0.221	0.406	0.367	0.361	0.250	0.224	0.392	0.344	0.556	0.444	0.436	0.327	0.628	0.239	0.453	0.247	0.252	0.252
25	0.425	0.294	0.294	0.303	0.294	0.531	0.294	0.294	0.330	0.294	0.319	0.531	0.531	0.286	0.411	0.342	0.330	0.339
26	0.250	0.303	0.222	0.244	0.241	0.428	0.283	0.227	0.328	0.330	0.381	0.353	0.275	0.176	0.367	0.258	0.241	0.252
27	0.431	0.517	0.517	0.517	0.408	0.367	0.517	0.517	0.431	0.431	0.431	0.353	0.422	0.361	0.422	0.336	0.302	0.289
28	0.425	0.327	0.327	0.327	0.286	0.575	0.327	0.300	0.330	0.330	0.327	0.327	0.575	0.569	0.294	0.411	0.342	0.330
29	0.417	0.336	0.336	0.336	0.294	0.417	0.336	0.336	0.336	0.336	0.336	0.417	0.417	0.289	0.411	0.342	0.330	0.339
30	0.450	0.428	0.403	0.428	0.250	0.380	0.450	0.442	0.450	0.450	0.450	0.380	0.352	0.294	0.352	0.380	0.339	0.369
31	0.536	0.417	0.403	0.467	0.389	0.464	0.428	0.417	0.489	0.536	0.536	0.425	0.461	0.361	0.453	0.481	0.481	0.481
32	0.375	0.317	0.317	0.317	0.317	0.433	0.317	0.317	0.330	0.330	0.381	0.375	0.375	0.317	0.444	0.489	0.489	0.489
33	0.536	0.428	0.403	0.467	0.331	0.464	0.533	0.442	0.536	0.536	0.536	0.425	0.492	0.331	0.453	0.389	0.339	0.369
34	0.425	0.403	0.356	0.419	0.294	0.642	0.419	0.300	0.419	0.419	0.419	0.864	0.569	0.294	0.411	0.389	0.369	0.369
35	0.442	0.456	0.456	0.456	0.408	0.322	0.456	0.456	0.442	0.442	0.442	0.322	0.291	0.369	0.369	0.336	0.302	0.280
36	0.394	0.406	0.394	0.394	0.361	0.325	0.394	0.394	0.556	0.444	0.436	0.327	0.628	0.361	0.453	0.336	0.319	0.319

	19	20	21	22	23	24	25	26	27	28	29	30	31	32	33	34	35	36
1	0.165	0.272	0.325	0.397	0.325	0.221	0.286	0.250	0.289	0.327	0.289	0.380	0.536	0.311	0.517	0.419	0.249	0.225
2	0.406	0.381	0.358	0.414	0.392	0.428	0.294	0.381	0.517	0.327	0.336	0.428	0.417	0.317	0.428	0.403	0.456	0.406
3	0.369	0.381	0.358	0.414	0.314	0.367	0.202	0.222	0.517	0.221	0.258	0.210	0.389	0.317	0.331	0.230	0.456	0.361
4	0.389	0.381	0.358	0.414	0.361	0.428	0.303	0.381	0.517	0.327	0.336	0.428	0.467	0.317	0.467	0.419	0.456	0.394
5	0.369	0.381	0.358	0.408	0.314	0.367	0.294	0.286	0.408	0.286	0.294	0.250	0.389	0.317	0.331	0.294	0.408	0.361
6	0.366	0.433	0.433	0.433	0.428	0.347	0.531	0.428	0.367	0.575	0.417	0.380	0.464	0.433	0.464	0.642	0.322	0.325
7	0.392	0.381	0.358	0.414	0.392	0.436	0.294	0.381	0.517	0.327	0.336	0.450	0.428	0.317	0.533	0.419	0.456	0.394
8	0.389	0.381	0.358	0.344	0.392	0.436	0.294	0.381	0.517	0.300	0.336	0.442	0.417	0.317	0.442	0.300	0.456	0.394
9	0.556	0.381	0.358	0.397	0.392	0.556	0.330	0.381	0.431	0.330	0.336	0.450	0.489	0.330	0.536	0.419	0.442	0.556
10	0.444	0.381	0.358	0.411	0.392	0.444	0.294	0.381	0.431	0.327	0.336	0.450	0.536	0.330	0.536	0.419	0.442	0.444
11	0.389	0.151	0.142	0.173	0.331	0.436	0.170	0.381	0.431	0.207	0.286	0.450	0.417	0.151	0.536	0.272	0.442	0.394
12	0.286	0.375	0.333	0.392	0.367	0.327	0.531	0.353	0.225	0.575	0.417	0.224	0.403	0.375	0.314	0.864	0.269	0.319
13	0.628	0.375	0.333	0.392	0.392	0.628	0.531	0.353	0.422	0.569	0.417	0.352	0.461	0.375	0.492	0.569	0.291	0.628
14	0.182	0.181	0.269	0.289	0.286	0.239	0.174	0.176	0.353	0.199	0.252	0.224	0.339	0.264	0.289	0.261	0.369	0.283
15	0.453	0.444	0.444	0.444	0.392	0.453	0.411	0.367	0.422	0.411	0.411	0.352	0.453	0.444	0.453	0.411	0.369	0.453
16	0.336	0.525	0.639	0.639	0.383	0.336	0.342	0.342	0.336	0.342	0.342	0.380	0.481	0.489	0.389	0.389	0.336	0.336
17	0.302	0.525	0.653	0.661	0.383	0.327	0.330	0.330	0.302	0.330	0.330	0.339	0.481	0.489	0.339	0.339	0.302	0.319
18	0.289	0.525	0.653	0.661	0.383	0.252	0.325	0.252	0.176	0.255	0.303	0.207	0.481	0.489	0.283	0.369	0.196	0.250
19	1.000	0.369	0.358	0.369	0.392	0.717	0.294	0.381	0.422	0.300	0.336	0.389	0.461	0.366	0.492	0.325	0.389	0.717
20	0.369	1.000	0.525	0.525	0.383	0.367	0.375	0.353	0.381	0.375	0.375	0.328	0.481	0.489	0.331	0.375	0.381	0.361
21	0.358	0.525	1.000	0.653	0.383	0.358	0.333	0.333	0.358	0.333	0.333	0.325	0.481	0.489	0.331	0.369	0.358	0.358
22	0.369	0.525	0.653	1.000	0.383	0.367	0.392	0.353	0.414	0.392	0.392	0.380	0.481	0.489	0.397	0.397	0.414	0.361
23	0.392	0.383	0.383	0.383	1.000	0.392	0.367	0.461	0.392	0.367	0.367	0.352	0.392	0.461	0.428	0.369	0.331	0.392
24	0.717	0.328	0.275	0.336	0.392	1.000	0.294	0.347	0.422	0.300	0.336	0.352	0.461	0.222	0.492	0.297	0.291	0.745
25	0.294	0.375	0.333	0.392	0.367	0.327	1.000	0.353	0.294	0.531	0.417	0.294	0.403	0.375	0.314	0.531	0.291	0.319
26	0.366	0.319	0.269	0.306	0.461	0.347	0.241	1.000	0.367	0.236	0.209	0.300	0.392	0.597	0.428	0.325	0.322	0.325
27	0.422	0.381	0.358	0.414	0.392	0.431	0.294	0.381	1.000	0.300	0.336	0.431	0.422	0.367	0.431	0.325	0.456	0.422
28	0.300	0.375	0.333	0.392	0.367	0.327	0.531	0.353	0.300	1.000	0.417	0.327	0.403	0.375	0.327	0.575	0.291	0.319
29	0.336	0.375	0.333	0.392	0.367	0.336	0.417	0.353	0.417	0.417	1.000	0.336	0.403	0.375	0.336	0.417	0.291	0.336
30	0.389	0.328	0.325	0.380	0.352	0.436	0.294	0.381	0.431	0.327	0.336	1.000	0.417	0.311	0.450	0.380	0.442	0.394
31	0.461	0.481	0.481	0.481	0.392	0.461	0.403	0.392	0.422	0.403	0.403	0.417	1.000	0.481	0.517	0.419	0.417	0.461
32	0.366	0.489	0.489	0.489	0.461	0.347	0.375	0.597	0.367	0.375	0.375	0.311	0.481	1.000	0.428	0.375	0.322	0.325
33	0.492	0.331	0.331	0.397	0.428	0.492	0.314	0.428	0.431	0.327	0.336	0.450	0.517	0.428	1.000	0.419	0.442	0.492
34	0.325	0.375	0.369	0.397	0.369	0.327	0.531	0.353	0.325	0.575	0.417	0.380	0.419	0.375	0.419	1.000	0.322	0.325
35	0.389	0.381	0.358	0.414	0.331	0.436	0.291	0.381	0.456	0.291	0.291	0.442	0.417	0.322	0.442	0.322	1.000	0.394
36	0.717	0.361	0.358	0.361	0.392	0.745	0.319	0.381	0.422	0.319	0.336	0.394	0.461	0.325	0.492	0.325	0.394	1.000

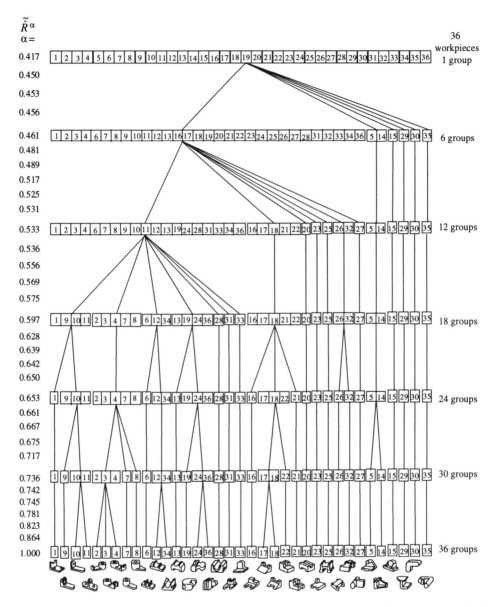

Fig. 12. Lean classification inferred from data for 8 typical workpieces using the max-min method and denoted as $C8$.

identical in the lean and benchmark classification systems. The idea is to take the classification results of the benchmark classification as a criterion and calculate the ratio of the sum of the number of workpieces in each corresponding group in the two classifications that are the same over the total number of workpieces within each group. This ratio will be referred to as the degree of consistency between the lean

Table 9. The two sets of classification results in Example 4.

1st Classification \ 2nd Classification	$B_1 = \{x_1, x_9\}$	$B_2 = \{x_2, x_3, x_4\}$	$B_3 = \{x_5, x_6, x_7, x_{10}\}$	$B_4 = \{x_8\}$	Number of Workpieces in Corresponding Groups	Workpiece(s) in Corresponding Groups
$A_1 = \{x_1\}$	①	0	0	0	1	x_1
$A_2 = \{x_2, x_4, x_5, x_6\}$	0	②	2	0	2	x_2, x_4
$A_3 = \{x_3\}$	0	1	0	⓪	0	
$A_4 = \{x_7, x_8, x_9, x_{10}\}$	1	0	②	1	2	x_7, x_{10}

and benchmark classification systems. The higher the degree of consistency, the more consistent the two sets of classification results. Let the term corresponding groups refer to the groups in the two classification schemes that have a one-to-one relationship when the classifications have the same number of groups. The algorithm used for measuring the degree of consistency between two classifications is shown in Example 4.

Example 4. Consider a set of 10 sample workpieces $(x_1, x_2, \ldots, x_{10})$. Suppose two classification schemes each include four groups, as shown in Table 9. We will use the above algorithm to calculate the degree of consistency between the two sets of classification results.

According to the algorithm mentioned in Sec. 3.2, we eventually obtain four pairs of corresponding groups: A_1 and B_1, A_2 and B_2, A_3 and B_4, and A_4 and B_3 for this example. The number of workpieces shared between each of these pairs of corresponding groups is 1, 2, 0, and 2, respectively. Thus $h = (1+2+0+2)/10 = 0.5$, which is the degree of consistency between these two classification schemes when the number of groups is four.

5.2. *Evaluating the effectiveness of lean classification*

The study employs the fuzzy clustering method to proceed workpiece classification. In either the lean classification or the benchmark classification, users' judgments concerning the degree of similarity between different workpieces can be employed to choose the number of groups used in the classification scheme.

Generally speaking, for practical application of GT in design and manufacturing, a middle-range value of α should usually be chosen, so that there are neither very many nor very few groups. In the example considered earlier, with 36 sample workpieces, suppose the user thinks there should be between 18 and 24 groups. In this case, the degree of consistency between the lean classification system built up from different numbers of typical workpieces (for example, $C_2, C_4, \ldots, C_{36}$) and the benchmark classification system is as shown in Fig. 13. The figure shows that if a lean classification system is formulated on the basis of 8 typical workpieces (C_8), with 24 groups, then the degree of consistency between the lean and benchmark

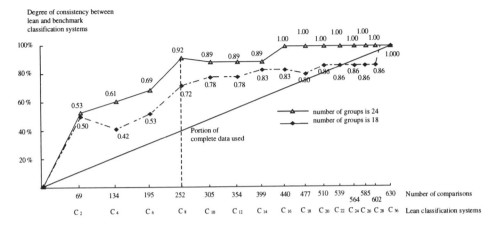

Fig. 13.   Degree of consistency for each lean classification with between 18 and 24 groups.

classification systems will be 92%. However, the cost of establishing the lean classification system using partial experimental data will be only 40% of that of establishing a classification with complete experimental data. (In this case, the number of partial experimental data for 8 typical workpieces is 252 while that for the complete experimental data is 630, 252/360 = 0.40.)

If we use the ratio of the degree of consistency over the proportion of the complete experimental data used in the lean classification system as a measure of the efficiency of the lean classification system, then for a classification system with 24 groups, if we accept the accuracy percentage of 92%, we can reach the same efficiency of the benchmark classification system using only 40% of the complete experimental data. The efficiency of the lean classification system is 2.3 times greater than that of the benchmark classification system. Another way of understanding the advantages of the lean classification is that users can first determine the required number of groups, determine the degree of consistency with the benchmark classification that they require, and then select the number of typical workpieces needed to produce a lean classification with the same degree of consistency. In this way, a lean classification system using a proper number of typical workpieces can be used to replace the benchmark classification system.

Figure 14 shows the process of calculating the average degree of consistency with the benchmark classification of lean classification systems (for example, $C_2, C_4, \ldots, C_{36}$) comprising different numbers of groups. The result can be seen in Fig. 15. In the case of the lean classification system with 8 typical workpieces (C_8), each of the clustering results from different hierarchies (Fig. 12) is compared with the result from the benchmark classification (Fig. 8) to calculate their degree of consistency. Finally, an average degree of consistency of 78% is reached. In other words, the classification results based on only 40% of the complete experimental data are 78% consistent with those of the benchmark classification system. Moreover, the lean classification is 1.95 times more efficient than the benchmark classification system.

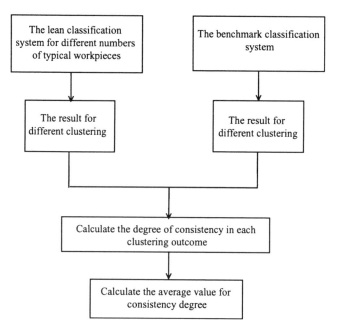

Fig. 14.   Calculating the average consistency degree between the lean classification and the bench-mark classification systems.

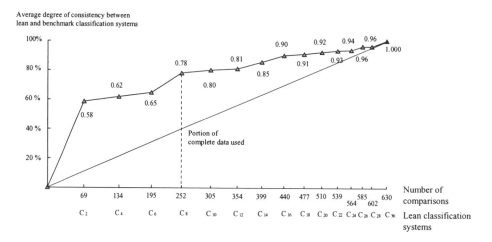

Fig. 15.   Average degree of consistency for lean classifications with different numbers of groups.

6. Simulation Method for Evaluating the Effectiveness and Efficiency of Lean Classification

The example used in Sec. 5 to demonstrate the efficiency of the lean classification system included only 36 sample workpieces. This chapter adopted a simulation method to determine if the lean classification system maintains a high level of efficiency, even when a large number of sample workpieces is classified. First of all, a matrix will be generated by the computer. The dimensions of the matrix are determined by the number of sample workpieces. The range of its elements is [0, 1] and features the symmetrical characteristics. Moreover, the value of the elements in the diagonal is set to 1. Such a matrix is used to simulate the similarity of the sample workpieces after the pair comparison. It is, therefore, defined as the simulation matrix. According to the procedure mentioned earlier, we will present a simulation method for estimating the efficiency of a lean classification system for sets of from 100 to 800 sample workpieces in this chapter. In the first subsection below, we construct a simulation matrix. In the second subsection, we determine a reasonable interval of degrees of similarity between two workpieces. In the third subsection, we simulate the efficiency of a lean classification system for a large number of workpieces.

6.1. *Constructing a simulation matrix*

To use a simulation method to evaluate the efficiency of lean classification, we need to construct a simulation matrix S. The elements in the matrix will denote the simulated degree of similarity as a subject performs pair comparisons of n workpieces $(x_1, x_2, \ldots, x_n)$. The elements of the matrix are interrelated in certain ways. For example, if both two workpieces are very similar to a third one, then these two workpieces will have high degree of similarity. Therefore, the values of the elements in S cannot be assigned randomly, but must be assigned in light of the interrelations between the elements. For convenience, we will utilize the reflexive and symmetric properties of the matrix, that is, $s_{ii} = 1$, $s_{ij} = s_{ji}$, and consider only the upper part of the data in our discussion.

$$S = [s_{ij}]_{nn} = \begin{array}{c} \\ x_1 \\ x_2 \\ x_3 \\ \vdots \\ x_n \end{array} \begin{array}{cccccc} x_1 & x_2 & x_3 & \cdots & x_n \\ \left(\begin{array}{ccccc} 1 & s_{12} & s_{13} & \cdots & s_{1n} \\ & 1 & s_{23} & \cdots & s_{2n} \\ & & 1 & \cdots & s_{3n} \\ & & & \ddots & \vdots \\ & & & & 1 \end{array}\right) \end{array}. \qquad (14)$$

The values of the elements in the simulation matrix are produced as follows.

Step 1. Assign preliminary values to the first row
Assign random numbers [0, 1] to the first row of the matrix s_{1j}, $j = 2, \ldots, n$; this means the degree of similarity between workpiece x_1 and other workpieces $(x_2, x_3, \ldots, x_n)$ is known.

Step 2. Determine a reasonable interval among other rows s_{ij}
From the degree of similarity between workpiece x_1 and other workpieces, a reasonable interval of degrees of similarity between any two workpieces, except those in

the first row, can be inferred; that is, we infer the interval that s_{ij}, $i = 2, 3, \ldots, n$ and $j = i + 1, \ldots, n$ may fall in (please refer to Sec. 6.2 for the detail).

Step 3. Assign s_{ij} a value in the interval determined in Step 2

Randomly assign s_{ij} a value in the interval of degrees of similarity between any two workpieces. This kind of simulation matrix will reflect that all the workpieces are interrelated in certain ways.

6.2. *Determining interval of degrees of similarity between two workpieces*

In constructing the simulation matrix, we need to know the degree of similarity between workpiece x_1 and other workpieces in order to infer a general interval for the degree of similarity between any two workpieces (except for the first row of elements in the matrix). The theoretical basis of the inference is to regard the information content of the global shape of each workpiece as a set. The total value is assumed to be 100%; that is, 1. The degree of similarity between the global shapes of two workpieces is then just the intersection of two sets. The inference method is to utilize the intersection of the sets representing the global shape information of any two workpieces and workpiece x_1 to determine the smallest and largest possible intersection between the two workpieces. These will then serve as the lower bound and upper bound of the interval of the degree of similarity between the two workpieces.

Take two workpieces x_p and x_q as an example. Let s_p, s_q and, s_1 denote the global shape information of the three workpieces x_p, x_q, and x_1, respectively. Assume that the degree of similarity between x_p and x_1 is s_{p1} and that between x_1 and x_q is s_{1q}, and we want to find a reasonable interval of the degree of similarity between the two workpieces x_p and x_q. First, we find the lower bound of the interval of s_{pq} by finding the smallest intersection between s_p, s_q and s_1 (see Fig. 16). When s_p and s_q are on different sides of s_1, then $s_{p1} = s_p \cap s_1$ and $s_{1q} = s_1 \cap s_q$; when $s_{p1} + s_{1q} > 1$, the lower bound of s_{pq} is $s_{p1} + s_{1q} - 1$ (Fig. 16(a)); and when $s_{p1} + s_{1q} \leq 1$, the lower bound of s_{pq} equals 0 (Fig. 16(b)). These two formulae can be combined into the general formula where the lower bound is $\max[0, s_{p1} + s_{1q} - 1]$. Second, to calculate the upper bound of the interval of s_{pq} we find the largest intersection set between s_p, s_q and s_1 (see Fig. 17). Now s_p and s_q are both on the same side of s_1. When $s_{p1} \leq s_{1q}$, the upper bound of s_{pq} is $s_{p1} + (1 - s_{1q})$ (Fig. 17(a)); when $s_{p1} \geq s_{1q}$, the upper bound is $s_{1q} + (1 - s_{p1})$ (Fig. 17(b)). These two formulae can be combined into the general formula $1 - |s_{p1} - s_{1q}|$. Thus a reasonable interval for the degree of similarity between any two workpieces through x_1 is $[\max[0, s_{p1} + s_{1q} - 1]$, $[1 - |s_{p1} - s_{1q}|]]$. In the following section, we again use an example to illustrate this inference method.

Construct a 5×5 simulation matrix as an example. The preliminary values s_{1j}, $j = 2, 3, 4, 5$ of the elements in the first row of the matrix are assigned randomly.

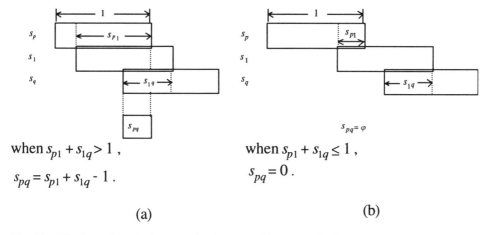

when $s_{p1} + s_{1q} > 1$,

$s_{pq} = s_{p1} + s_{1q} - 1$.

when $s_{p1} + s_{1q} \leq 1$,

$s_{pq} = 0$.

(a)

(b)

Fig. 16.   The lower bound of s_{pq} can be determined by using the degree of similarity with workpiece x_1.

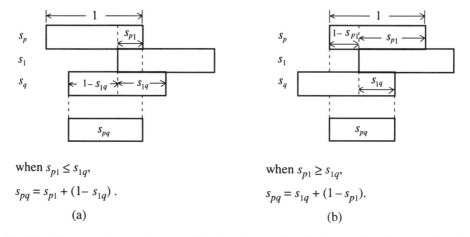

when $s_{p1} \leq s_{1q}$,

$s_{pq} = s_{p1} + (1 - s_{1q})$.

(a)

when $s_{p1} \geq s_{1q}$,

$s_{pq} = s_{1q} + (1 - s_{p1})$.

(b)

Fig. 17.   The upper bound of s_{pq} can be determined using the degree of similarity with workpiece x_1.

Now we infer the interval in which s_{ij}, $i = 2, 3, 4, 5$ and $j = 3, 4, 5$ may fall as follows.

$$
\begin{array}{c}
\begin{array}{ccccc}
x_1 & x_2 & x_3 & x_4 & x_5
\end{array} \\
\begin{array}{c}
x_1 \\ x_2 \\ x_3 \\ x_4 \\ x_5
\end{array}
\left(
\begin{array}{ccccc}
1 & 0.9 & 0.7 & 0.3 & 0.1 \\
 & 1 & [0.6, 0.8] & [0.2, 0.4] & [0, 0.2] \\
 & & 1 & [0, 0.6] & [0, 0.4] \\
 & & & 1 & [0, 0.8] \\
 & & & & 1
\end{array}
\right).
\end{array}
\tag{15}
$$

Take s_{23} as an example. Because the similarity level between x_2 and x_1 is 0.9 and that between x_3 and x_1 is 0.7, the lower bound of s_{23} is $\max[0, 0.9 + 0.7 - 1] = 0.6$

and the upper bound is $1 - |0.9 - 0.7| = 0.8$. Therefore, a reasonable interval of the degree of similarity between x_2 and x_3 can be inferred to be $[0.6, 0.8]$. Values are assigned randomly within this interval to construct the simulation matrix.

6.3. *Using a large sample to examine the efficiency of lean classification*

Once the simulation matrix has been constructed, similar to the experimental method mentioned in Sec. 2, we can formulate a simulated benchmark classification on the basis of all the data in the matrix. This can be defined as a pseudo-benchmark classification. We can also take a portion of the data and use the max-min method to infer classification results for the complete data, thus obtaining a simulated lean classification, which can be defined as a pseudo-lean classification. By measuring the consistency between the two classification results and comparing the cost of establishing the lean classification and benchmark classification, we can determine the efficiency of the pseudo-lean classification. As it is very difficult to judge the global shapes of n sample workpieces from the simulation matrix, the typical work-pieces used in the pseudo-lean classification are selected not by the binary cluster-ing method described in Sec. 4, but directly from the n sample workpieces by their order, based on the required number of typical workpieces. For instance, to establish a pseudo-lean classification base on six typical workpieces, we can take the first six rows of elements directly from the simulation matrix as partial experimental data.

To examine the efficiency of the lean classification method, we constructed four simulation matrixes, with 100, 200, 400, and 800 sample workpieces, respectively. The values of the elements were set to the second decimal place, and then the lean classification was simulated with different typical workpieces. To compare the classification results for these four sets of sample workpieces, we will use pseudo-lean classifications employing from 10% to 100% of the complete experimental data to calculate the average degree of consistency with the benchmark classification for different numbers of groups (see Table 10).

From Table 10, we can make three interesting observations. First, the results of lean classification based on only a small portion of the data are still highly con-sistent with the results of benchmark classification. For example, for 800 sample workpieces, the degree of consistency between the lean classification based on only 10% of the complete data and the benchmark classification is 71%. Second, when the percentage of the experimental data used in formulating the lean classification is doubled, only a limited increase occurs in the degree of consistency with the benchmark classification. For instance, suppose there are 800 sample workpieces. If the percentage of the experimental data used is increased fivefold; that is, if a lean classification is established with 50% of the complete data, then the degree of consistency between the lean and benchmark classifications will be 78%, an increase of only 7%. Moreover, the size of this increase continues to shrink as the amount of

Table 10. Degree of consistency between pseudo-lean and benchmark classifications for samples of 100 to 800 workpieces.

No. of sample workpieces \ % experimental data	10%	20%	30%	40%	50%	60%	70%	80%	90%	100%	Remarks
100	0.65	0.67	0.68	0.74	0.77	0.80	0.80	0.83	0.90	1.00	Using simulation matrix instead of pair comparison data, directly selecting samples to establish lean classification
200	0.67	0.67	0.68	0.72	0.75	0.77	0.83	0.85	0.91	1.00	
400	0.68	0.69	0.72	0.75	0.75	0.80	0.82	0.85	0.90	1.00	
800	0.71	0.71	0.73	0.76	0.78	0.79	0.84	0.87	0.91	1.00	

(Average degree of consistency between pseudo-lean/benchmark classification)

data used in the lean classification increases. Third, as the number of sample work-pieces increases, a lean classification using the same percentage of the data achieves an increasing degree of consistency with the benchmark classification. For exam-ple, suppose the number of sample workpieces increases from 100 to 800. Then the degree of consistency between the benchmark classification and a lean classification based on just 10% of the complete experimental data increases from 65% to 71%.

These simulation results show that in a group technology practical workpiece classification application, as the number of sample workpieces increases, only a small portion of the complete experimental data is needed to produce a lean classification that is highly consistent with the benchmark classification. For a sample of 800 workpieces, for example, gathering complete experimental data would involve mak-ing 319,600 pair comparisons. However, if the lean classification is used instead, we can produce a classification that is 71% consistent with the benchmark classification while using just 40 typical workpieces (less than 10% of the total data) and making only 31,180 pair comparisons.

7. Conclusions and Future Work Suggestions

7.1. *Conclusions*

To evaluate the performance of an automatic classification system, we need to exam-ine whether its classification results are consistent with users' judgments. The higher the degree of consistency, the more effective the classification. Therefore, classifica-tion benchmarks based on users' judgments are necessary. The proposed evaluation technique for the effectiveness of the automated workpiece classification systems boasts two salient characteristics. On the one hand, it is user-oriented. That is to say, the benchmark classification is based on users' judgements of the degree of similarity between samples in terms of the global shape. It is, therefore, feasible to verify whether or not the classification system is compatible with the user's mental model. On the other hand, it allows a flexible similarity criteria to be applied, i.e. different workpiece classification benchmarks can be built up in accordance with the user's point of view toward the similarity of the workpieces. This technique makes it possible for users to choose a practical automated classification system to classify the general workpiece population. It establishes a clustering database for workpieces of varying similarity in accordance with their information retrieval needs.

A benchmark classification provides an accurate evaluation of the performance of an automatic classification system, but when a large number of samples are involved, it may be costly, time-consuming, and unreliable because of bias due to fatigue among the subjects.

To cope with these shortcomings, the authors have proposed a lean classification method, in which only a relatively small number of typical workpieces are used to make pair comparisons with all of the sample workpieces. The partial experimental data are then used to infer results similar to those that would be obtained from the complete experimental data. In an experiment with a small set of 36 sample workpieces divided among 24 classification groups, the results of a lean classification established using 40% of the complete experimental data were 92% consistent with those of the benchmark classification. The efficiency of the lean classification is 2.3 times higher. If we take the average degree of consistency between the classification results for various numbers of classification groups, the lean classification is 78% consistent with the benchmark classification and 1.95 times more efficient. In simulations with medium to large samples of 100 to 800 workpieces, we found that when the number of workpieces increases, if we take the average degree of consistency of the classification results for various numbers of groups, a lean classification established with only 10% of the complete experimental data is 71% consistent with the benchmark classification, providing a 7.2 times increase in efficiency. This indicates that choosing a proper lean classification system can serve as a classification specification for assessing the effectiveness of an automated workpiece classification system. More importantly, it is efficient. The lean classification, however, has some limitations as far as its applicability is concerned, which needs to be further studied.

7.2. *Limitations of the application of lean classification*

To explore the limitations of the application of lean classification, first we need to know the performance interval of the automatic classification system in which the lean classification will be used to infer. The method was described in Sec. 6. In Fig. 18, for example, let a represent the degree of consistency between the lean classification and the benchmark classification. The consistency degree can be estimated using the simulation or heuristic method. Let b represent the degree of consistency between the results of the automatic classification system and those of the lean classification. b can also be regarded as the representation of the performance of the automatic classification system, as measured by the lean classification system. c represents the performance interval of the automated classification that is inferred by the lean classification. Given a and b, we want to infer the interval of c, the value of an interval by $[x, y]$, in which the lower boundary $x = \max[0, a + b - 1]$ and

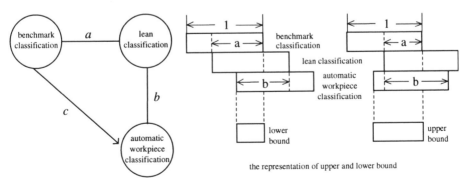

Fig. 18.   Using the lean classification to evaluate the performance of an automatic classification system with respect to benchmark classification.

the upper boundary $y = [1 - |a - b|]$. In an ideal situation, we can expect that the interval of c is very close to b. In such a condition, the lean classification system can substitute for the benchmark classification, which can be often seen when the value of a is very high.

Take C_8 (the lean classification of 8 typical workpieces) for an instance. The degree of consistency between C_8 and the benchmark classification is 92% when there are 24 clusterings. That is to say, $a = 0.92$. Comparing this lean classification with the automated classification system developed by Wu and Jen (the number of clustering is 24), the degree of consistency is 64%; i.e. $b = 0.64$. If there is no benchmark classification available, then we can infer that the automated classification developed by Wu and Jen should fall in the interval $[0.56, 0.72]$ compared with the benchmark classification. If it is accepted by the user, then it can be considered as a practical automated classification system. Otherwise, it will be rejected. Compared with the benchmark classification mentioned in the study, the consistency index of Wu and Jen's automated classification is 0.64, which is very close to the interval inferred. This demonstrates that the lean classification can take place of the benchmark classification when the value of a is very large.

The application of the lean classification system, however, is limited in the following situations:

Case 1. The value of a is low even though that of b is very high. For example, if $a = 0.40$, $b = 0.92$, then the consistency interval inferred by the automated classification system can reach $[0.32, 0.48]$, which is remarkably different from b. Obviously, the user will not accept that it is a practical automated classification system.

Case 2. When the value of b is very low. For example, $b = 0.40$. No matter what value a is assigned, the user will not consider it a practical automated classification system.

Case 3. Both a and b are not very high. For instance, if a equal to 0.71, a stimulated lean classification by 10% partial experimental data and b is 0.64, then the consistency interval of the automated classification system will fall in $[0.35, 0.93]$. Suppose the distribution pattern in the interval is a normal one and there are three standard deviations both to the left and to the right, the probability is only 4.9% (Fig. 19(a)) if the consistency index is set to 0.80 for an effective automated classification system. And if we assume that the distribution is a uniform pattern, then the probability will only reach 22.4% for the same consistency criteria (Fig. 19(b)). Because the probabilities in both situations are very low, the user will not take the risk to accept such an automated classification system in these consistency intervals.

In these three cases, the lean classification system is unable to evaluate whether an automated classification is feasible. The authors, therefore, would not recommend the lean classification system application because it might not be effective in terms

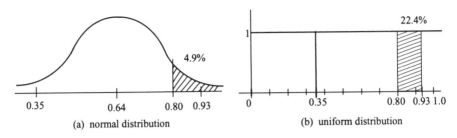

Fig. 19. The probability high than 0.8 in the interval of [0.35 , 0.93].

of workpiece classification. In other words, the lean classification can function as a tool for us to recognize automated workpiece classifications that are not efficient or economical. Hence in this case, the lean classification would only be useful for sifting out ineffective automated classification systems in the beginning.

7.3. *Future work suggestions*

In this study, the author first proposed that the benchmark classification based upon the user's mental model can serve as the foundation for the evaluation of automated classification systems. Furthermore, the lean classification can take place of the benchmark classification if it is verified by a simulation or heuristic method. Under the framework, the following issues are in need of further investigation:

(1) Explore the mechanism and strategy the expert and novice adopts in workpiece classification. The addition of the mechanism and strategy the expert uses in workpiece classification should be helpful in the training of a novice.
(2) In the evaluation of effectiveness of the lean classification, the similarity distribution pattern is hypothesized to be uniform, falling in the range of [0, 1]. Moreover, it was not verified whether or not the max-min method we used to infer the unknown data structure can reflect the highest degree of consistency between the simulated lean classification system and the stimulated benchmark classification system. Different types of similarity distribution patterns of workpieces and different inference methods should be examined.

Acknowledgement

The authors would like to thank the employees of the Aircraft Manufactory, Aerospace Industrial Development Corporation (AIDC), Taichung, Taiwan, ROC, for their participation in the workpiece pair comparison experiment.

Appendix

Four major concepts from fuzzy set theory were used in this research — fuzzy sets, fuzzy numbers, linguistic variables, and the α-cut of the membership matrix.

A1. *Fuzzy sets*

Fuzzy set theory was introduced by Zadeh.[24] It can be used to deal with problems in which a source of vagueness is present. It can be considered a modeling language that approximates situations in which fuzzy phenomena and criteria exist. Consider a reference set X with x as its element. A fuzzy subset $\tilde{A}$ of x is defined by a membership function $\mu_{\tilde{A}(x)}$ which maps each element x in X to a real number in the interval $[0, 1]$. The value of $\mu_{\tilde{A}(x)}$ denotes the grade of membership, that is, the degree to which element x is a member of set $\tilde{A}$. A fuzzy subset is often referred to briefly as a fuzzy set.[25]

A2. *Fuzzy numbers*

A fuzzy number is a special fuzzy subset in R (real line) which is usually represented by a special membership function over a closed interval of real numbers. In this research, a special class of fuzzy numbers known as trapezoidal fuzzy numbers (TrFN) developed by Jain[12] and Dubois and Prade[9] was used. As shown in Fig. A.1, a TrFN has a trapezoidal shape and can be denoted by (a, b, c, d), where a can be semantically interpreted as the lower bound, b and c are the most probable value, and d is the upper bound, with the membership function defined as follows.

$$\mu_{\tilde{A}(x)} = \begin{cases} 0 & (x \leq a) \\ \dfrac{x - a}{b - a} & (a < x < b) \\ 1 & (b \leq x \leq c) \\ \dfrac{x - d}{c - d} & (c < x < d) \\ 0 & (x \geq d). \end{cases}$$

Using the extension principle proposed by Zadeh,[24] the addition and subtraction operations on TrFNs definitely yield a TrFN. Multiplication, inverse, and division operations on TrFNs do not necessarily yield a TrFN. However, the results of these operations can be reasonably approximated by TrFNs,[14] as illustrated below.
Addition $\oplus$

$$\tilde{A}_1 \oplus \tilde{A}_2 = (a_1, b_1, c_1, d_1) \oplus (a_2, b_2, c_2, d_2)$$
$$= (a_1 + a_2, b_1 + b_2, c_1 + c_2, d_1 + d_2).$$

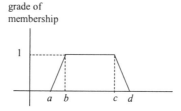

Fig. A.1.   Membership function of a trapezoidal fuzzy number $\tilde{A} = (a, b, c, d)$.

Subtraction $\ominus$

$$\tilde{A}_1 \ominus \tilde{A}_2 = (a_1, b_1, c_1, d_1) \ominus (a_2, b_2, c_2, d_2)$$
$$= (a_1 - d_2, b_1 - c_2, c_1 - b_2, d_1 - a_2).$$

Multiplication $\otimes$

$$k \otimes \tilde{A}_1 = k \otimes (a_1, b_1, c_1, d_1)$$
$$= (ka_1, kb_1, kc_1, kd_1) \qquad \text{if } k \geq 0$$
$$\tilde{A}_1 \otimes \tilde{A}_2 = (a_1, b_1, c_1, d_1) \otimes (a_2, b_2, c_2, d_2)$$
$$\cong (a_1 a_2, b_1 b_2, c_1 c_2, d_1 d_2) \qquad \text{if } a_1 \geq 0, a_2 \geq 0.$$

Division $\oslash$

$$\tilde{A}_1 \oslash \tilde{A}_2 = (a_1, b_1, c_1, d_1) \oslash (a_2, b_2, c_2, d_2)$$
$$\cong (a_1/d_2, b_1/c_2, c_1/b_2, d_1/a_2) \qquad \text{if } a_1 \geq 0, a_2 > 0.$$

In the special case when $b = c$, the trapezoidal fuzzy number (a, b, c, d) equals a triangular fuzzy number. The extended algebraic operations on triangular fuzzy numbers are the same as those on trapezoidal numbers.

A3. *Linguistic variables*

Linguistic variables are variables whose values are represented in words or sentences in natural languages. Each linguistic value can be modeled by a fuzzy set.[24] For example, let $\tilde{S}$ be a linguistic variable with the name "similarity" (the pair comparison between any two workpieces), and let the set of its linguistic terms be {very low similarity, low similarity, medium similarity, high similarity, very high similarity}. Each of these linguistic terms can be represented by a TrFN with its membership functions, as shown in Fig. A.2. Note that these linguistic terms are trapezoidal fuzzy numbers in the interval $[0, 1]$.

From the figure, the membership function of "low" is $(0.1, 0.25, 0.25, 0.4)$. That is, an expression of "low similarity" between two workpieces is between 0.1 and 0.4, and the most probable value is 0.25. The membership function for the degree of similarity at 0.2 in low is 0.67, while for the degree of similarity at 0.35 in low it is 0.33. Linguistic variables are useful for allowing experts to express uncertain judgments, such as those concerning the workpiece pair comparisons.

A4. *α-cut of membership matrix*

Let X be the universe of discourse, $X = \{x_1, x_2, \ldots, x_n\}$. Suppose that $\tilde{A}$ is a fuzzy number with membership function $\mu_{\tilde{A}}$. Then for every $\alpha \in [0, 1]$, the set $\tilde{R}^\alpha = \{x | \mu_{\tilde{A}}(x) \geq \alpha\}$ is called an α-cut of $\tilde{A}$.[25]

Similarly, suppose $\tilde{R}$ is a fuzzy relation, $\tilde{R} = [r_{ij}]_{m \times n}$, for every $\alpha \in [0, 1]$, $\tilde{R}^\alpha = [r_{ij}^\alpha]$ is called an α-cut of the membership matrix where

$$r_{ij}^\alpha = \begin{cases} 1 & \text{if } r_{ij} \geq \alpha \\ 0 & \text{if } r_{ij} < \alpha. \end{cases}$$

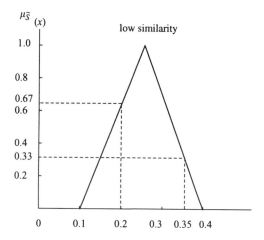

Fig. A.2.   The membership function for linguistic term "low similarity".

Terminology and Definition

C : Benchmark classification system.

C_x : A lean classification system built up by x typical workpieces, for example, C_8 means a lean classification system made up of 8 typical workpieces.

h : Index of the degree of consistency between two classifications.

r : Index of appropriate number of workpiece groups.

$\tilde{R}$: Complete experimental data which can be looked upon as a fuzzy relation.

$\hat{\tilde{R}}$: When $\tilde{R}$ satisfies the properties of reflexive, symmetric, and the transitive characteristics of $\tilde{R} \circ \tilde{R} \subseteq \tilde{R}$, it representative a similarity relation.

$\underline{\tilde{R}}_x$: Quasi-complete experimental data inferred by x representative workpieces; for example, $\underline{\tilde{R}}_8$ means the Quasi-complete experimental data inferred by 8 typical workpieces.

$\tilde{R}^\alpha$: α-cut of membership matrix of $\tilde{R}$.

$\tilde{S}_{ij}$: Represents the membership function after aggregation of the n subjects' similarity comparison between workpiece i and workpiece j, where $i < j$.

References

1. R. Arnheim, *The Power of Center* (University of California Press, Ltd., Berkeley and Los Angeles, 1988).
2. H. Bandemer and W. Nather, *Fuzzy Data Analysis* (Kluwer Academic Publishers, 1992).
3. A. Bhadra and G. W. Fischer, A new GT classification approach: A data base with graphical dimensions, *Manufacturing Review* **11** (1988) 44–49.
4. F. Bouslama and A. Ichikawa, Fuzzy control rules and their natural control laws, *Fuzzy Sets and Systems* **48** (1992) 65–86.
5. J. J. Buckley, The multiple judge, multiple criteria ranking problem: A fuzzy set approach, *Fuzzy Sets and Systems* **13** (1984) 25–37.

6. C. S. Chen, A form feature oriented coding scheme, *Computers and Industrial Engineering* **17** (1989) 227–233.

7. S. J. Chen and C. L. Hwang, *Fuzzy Multiple Attribute Decision Making — Method and Application, A State-of-the-Art Survey* (Springer-Verlag, New York, 1992).

8. L. A. Cooper, Mental representation of three-dimensional objects in visual problem solving and recognition, *Journal of Experimental Psychology: Learning, Memory, and Cognition* **16** (1990) 1097–1106.

9. D. Dubois and H. Prade, Operations on fuzzy numbers, *International Journal of System Sciences* **9** (1978) 613–626.

10. M. R. Henderson and S. Musti, Automated group technology part coding from a three-dimensional CAD data-base, *Transactions of the ASME Journal of Engineering for Industry* **110** (1988) 278–287.

11. S. H. Hsu, T. C. Hsia and M. C. Wu, A flexible classification method for evaluating the utility of automated workpiece classification system, *International Journal of Advanced Manufacturing Technology* **13** (1997) 637–648.

12. R. Jain, Decision-making in the presence of fuzzy variables, *IEEE Transactions of Systems Man Cybernetics* **6** (1976) 698–703.

13. S. Kaparthi and N. Suresh, A neural network system for shape-based classification and coding of rotational parts, *International Journal of Production Research* **29** (1991) 1771–1784.

14. A. Kaufmann and M. M. Gupta, *Introduction to Fuzzy Arithmetic: Theory and Applications* (Van Nostrand Reinhold, New York, 1985).

15. G. J. Klir and T. A. Folger, *Fuzzy Set, Uncertainty, and Information* (Prentice-Hall, 1992).

16. W. Labov, The boundaries of words and their meanings, eds. C.-J. N. Bailey, and R. W. Shuy, *New Ways of Analyzing Variations English* (Georgetown University Press, Washington, DC, 1973).

17. T. Lenau and L. Mu, Features in integrated modelling of products and their production, *International Journal of Computer Integrated Manufacturing* **6**, 1–2 (1993) 65–73.

18. S. Murakami, H. Maeda and S. Imamura, Fuzzy decision analysis on the development of centralized regional energy control system, preprints IFAC Conference on *Fuzzy Information, Knowledge Representation and Decision Analysis*, 1983, 353–358.

19. S. K. Tan, H. H. Teh and P. Z. Wang, Sequential representation of fuzzy similarity relations, *Fuzzy Sets and Systems* **67** (1994) 181–189.

20. S. K. Thompson, *Sampling* (John & Sons, Inc., New York, 1992).

21. M. C. Wu and J. R. Chen, A skeleton approach to modelling 2D workpieces, *Journal of Design and Manufacturing* **4** (1994) 229–243.

22. M. C. Wu, J. R. Chen and S. R. Jen, Global shape information modelling and classification of 2D workpieces, *International Journal of Computer Integrated Manufacturing* **7**, 5 (1994) 216–275.

23. M. C. Wu and S. R. Jen, A neural network approach to the classification of 3D prismatic parts, *International Journal of Advanced Manufacturing Technology* **11** (1996) 325–335.

24. L. A. Zadeh, The concept of a linguistic variable and its application to approximate reasoning, Parts 1, 2 and 3, *Information Science* **8** (1975) 199–249, 301–357; **9** (1976) 43–58.

25. H. J. Zimmermann, *Fuzzy Set Theory and Its Application*, 2nd edn., (Kluwer Academic Publishers, 1991).

CHAPTER 6

COMPUTER METHODS AND APPLICATIONS FOR THE REDUCTION OF MACHINING SURFACE ERRORS IN MANUFACTURING SYSTEMS

M. Y. YANG and J. G. CHOI

Department of Mechanical Eng.,
Kasit, 373-1, Kusong-dong, Yusong-gu,
Taejon 305-701, South Korea
Tel.: +82-42-869-2114/5114

This chapter presents two highly effective significant techniques for reducing machining errors. The significance of these techniques rests on the fact that much of the manufacturing systems is involved with metal removal, i.e. machining and the very high accuracy that is normally required.

Keywords: Machining surface accuracy; CNC (Computer Numerical Control) techniques; active control methods.

1. Introduction

In manufacturing, machining processes are used when higher dimensional accuracy and higher surface finish are required. Moreover, the flexible nature of machining processes enables the cost-effective, fast manufacture of products in small batch sized production. However, in reality, the machining process is usually not perfect and they cause a machining error. Therefore, the machining error, which denote the difference between the desired surface and the actual surface, should be minimized efficiently. To improve the machining surface accuracy in manufacturing environments, methods to reduce or compensate for the machining errors are continuously being sought. For machining CNC machine tools, machining error sources can be classified into three categories. First, controller and drive mechanisms; second, mechanical deficiencies such as backlash, non-straightness of the ballscrews and spindle run-out; and third, cutting process effects such as tool deflection due to cutting force, tool wear, and chatter vibration. The first set of error sources can be reduced by the servo controller that is implemented in the CNC unit. Many advanced servo control algorithms, such as feedback control, feedforward control, cross coupling control etc., have been developed to reduce the errors that can be measured in real time by the feedback devices such as encorders and/or linear scales. Most of the second set of error sources such as backlash or lead screw error can also be reduced easily through compensation techniques in the CNC Unit.

However, the errors due to the third sets of sources cannot be reduced by the CNC controllers, since these errors are not closed in the servo control loops. Especially, it is very difficult to obtain a high degree of accuracy when using a cutting tool with a low stiffness, such as in endmill operations, since machining accuracy is mainly dependent on the stiffness of the cutting tool. A number of solutions have been suggested in reducing the machining error caused by the tool deflection. They can be classified into two groups. One of the most common methods of reducing this machining error is to adjust the feed rate, which can be set by a constrained cutting force or machining error.[1–9] This technique can be called as the feedrate adjustment method. A disadvantage of this method is that, by reducing the feedrate in order to lower the cutting forces, the tool may be operating at a level far below its potential thus making this method inefficient. Also frequent changes of the feedrate may result in a lower surface quality since the surface roughness is proportional to the feedrate. In contrast to this approach, an active control system, in which the tool position is shifted, can be suggested as a means of reducing the surface error.[10,11] The shifting of the tool is achieved by a specially designed controller and is effective in reducing the surface error. This technique can be called as the active control method. In this chapter, two examples from these two techniques will be explained respectively in detail.

2. Feedrate Adjustment Technique

2.1. *Cutting force calculation*

In the end milling process, the lateral cutting force exerted on the tool causes the tool to deflect because of the cantilever form of the tool and the lack of stiffness. The instantaneous tool deflection at a given angular position is affected by not only the magnitude of the instantaneous cutting force, but its distribution along the tool axis. The cutting force in the end milling can be analyzed by considering the chip load on the cutting edges as follows.[12–18]

In cutting using an end mill with a helix angle of α_h, the instantaneous cutting volume by the individual flute varies largely according to the axial position of the cutting edge as well as its angular position. In order to reflect the difference of cutting loads properly, it is necessary to divide the end mill into disc elements of thin thickness as shown in Fig. 1.

Considering a single disc element, for the case when the rotating speed of the cut is fast enough compared to its feedrate, such that the movement of the cutting edge can be approximated by the circular locus, the chip thickness t_c produced by the cutting edge located at an engagement angle α from the reference position is approximated as follows:

$$t_c = f_t \sin \alpha \tag{1}$$

where f_t is the feed per flute. Since the cutting force can be calculated by the product of the cutting volume and specific cutting stiffness, the instantaneous resultant

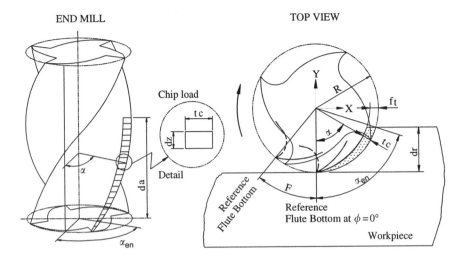

Fig. 1.   Chip load distribution model in end milling.

cutting force is represented as follows,

$$F = \sum_{k=1}^{N_t} \int_{d_{L,k}}^{d_{U,k}} K t_c \, dz \qquad (2)$$

where N_t is the number of flutes of the end mill and K is the specific cutting stiffness of work material. Also, $d_{U,k}$ and $d_{L,k}$ are the upper and the lower axial limits of chip load for the kth flute respectively, and have values between zero and the axial depth of cut d_a. Since an axial position on the cutter from its end z is related to the angular position α by following Eq. (3) with the cutter radius R, if the integral parameter is replaced with the angular position, the instantaneous cutting forces can be expressed by Eq. (4).

$$z = \frac{R}{\tan \alpha_h} \alpha \qquad (3)$$

$$F = \sum_{k=1}^{N_t} \int_{\alpha_{L,k}}^{\alpha_{U,k}} K \frac{R}{\tan \alpha_h} f_t \sin \alpha \, d\alpha$$

$$= K \frac{R}{\tan \alpha_h} f_t A(\alpha) \qquad (4)$$

Here, $\alpha_{U,k}$ and $\alpha_{L,k}$, shown in Fig. 2, are the upper and the lower angular limits of chip load for the kth flute, respectively, and $A(\alpha)$ is the chip load function defined as follows:

$$A(\alpha) = \sum_{k=1}^{N_t} [\cos \alpha_{L,k} - \cos \alpha_{U,k}] \qquad (5)$$

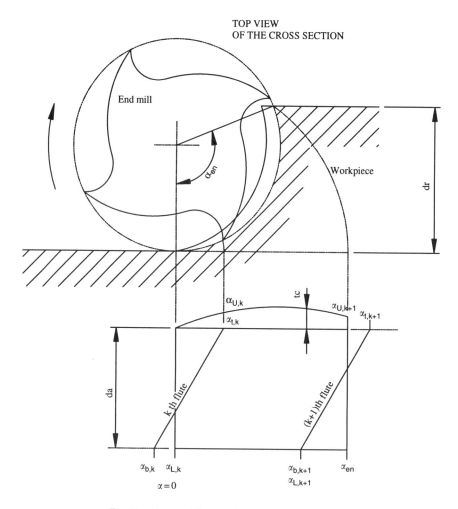

Fig. 2.   Engaged flute and chip load sweep angle.

Also, $\alpha_{t,k}$ and $\alpha_{b,k}$, which represent the cutter rotation angles at the top and the bottom limits by the kth flute in the cutting region, respectively, are calculated by Eqs. (6) and (7). From these, $\alpha_{U,k}$ and $\alpha_{L,k}$ in Eq. (5) can be obtained from Eqs. (8) and (9), respectively,

$$\alpha_{b,k} = (k-1)\frac{2\pi}{N_t} - \phi \tag{6}$$

$$\alpha_{t,k} = (k-1)\frac{2\pi}{N_t} - \phi + \beta \tag{7}$$

$$\alpha_{L,k} = \begin{cases} 0 & \text{when } -\beta \leq \alpha_{b,k} < 0 \\ \alpha_{b,k} & 0 \leq \alpha_{b,k} < \alpha_{en} \\ \emptyset & \text{otherwise} \end{cases} \tag{8}$$

$$\alpha_{U,k} = \begin{cases} \alpha_{t,k} & \text{when } 0 \leq \alpha_{t,k} < \alpha_{en} \\ \alpha_{en} & \alpha_{en} \leq \alpha_{t,k} < \alpha_{en} + \beta \\ \emptyset & \text{otherwise} \end{cases} \tag{9}$$

where α_{en} and β are the cutter engagement angle and the cutter sweep angle, respectively, defined as follows:

$$\alpha_{en} = \cos^{-1}\left(1 - \frac{d_r}{R}\right) \tag{10}$$

$$\beta = \frac{d_a}{R} \tan \alpha_h \tag{11}$$

In the expression for the instantaneous resultant cutting force in Eq. (4), the only variable is the chip load function. Accordingly, observation of the chip load function makes it possible to predict the change in cutting force. The chip load depends on the axial depth of cut d_a, the radial depth of cut d_r, and the cutter rotation angle ϕ. Figure 3 shows the expended chip load area and the engaged flutes.

The figure shows that a flute begins cutting at point a and finishes at point q of the chip load area. The flute engagement angle ϕ_e, i.e. the cutter rotation angle

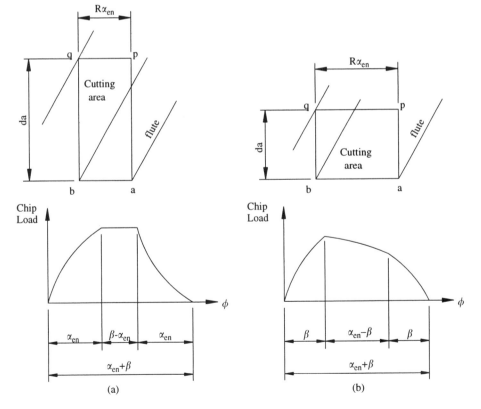

Fig. 3.   Cutting load area swept by cutting flutes.

during engagement by one flute is given as follows.

$$\phi_e = \frac{d_a}{R} \sin \alpha_h + \cos^{-1}\left(1 - \frac{d_r}{R}\right)$$
$$= \beta + \alpha_{en} \tag{12}$$

When this value is larger than the angle between neighboring flutes, two or more flutes simultaneously participate in cutting. Accordingly, in common cutting by an end mill with two or four flutes, two or more flutes may participate in cutting at the same time according to the magnitudes of the axial and the radial depths of cut.

Since the end milling is an interrupt cutting process where the cutting edges are located equally on the circumference of the cutter, the cutting force repeats periodically. The magnitude and the pattern of cutting force vary depending on cutting conditions. While the magnitude of cutting forces is directly influenced by the work material, cutting depth, feedrate, tool wear, and so on, the pattern of cutting force, which denotes the change of cutting force with time, depends predominantly on the tool shape and the cutting depth. The instantaneous tool deflection at a given angular position is affected by not only the magnitude of the instantaneous cutting force, but its distribution along the tool axis. Figure 4 shows that, while an end mill with equally spaced helical flutes rotates, the chip loads change according to its angular position. The variation of the chip loads induces the change of the magnitude and distribution of the cutting force and thus changes the force center, which is defined to be the location of application of the total point force necessary to produce the same bending moment about the fixed end of the tool as the distributed force. Accordingly, the tool deflection at each angular position also varies along with rotation of the tool.

2.2. *Prediction of surface error due to tool deflection*

During the rotation of the end mill, a flute sweeps from the bottom to the top of the machining surface and produces a new surface profile corresponding to the given axial depth. Since the tool deflection is continuously changing during the rotation, the surface profile does not coincide with an deflected tool shape, i.e. an instantaneous tool deflection generates only one point on the surface profile.[19–22]

The machined surface profile, which is the cross section perpendicular to the feed motion of the cutter, can be obtained from knowledge of the tool deflection and the axial position of the flute on the machined surface during successive cutter rotations. The tool deflection is closely related with the cutting force, and thus can be calculated easily by elementary beam theory if the elastic modulus of the cutter is known. The axial position of a flute on the machined surface can be determined by the angular position of a reference flute. Then, in order to obtain the surface error from the machined surface profile, the range of calculation in the axial direction of the cutter must be confirmed, which requires the axial depth of cut.

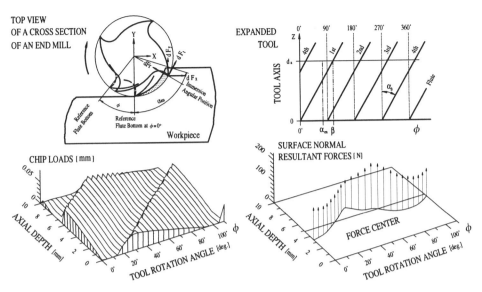

Fig. 4. Chip load distributions and cutting force changes in the cutting with an end mill of 6 mm diameter having 4 flutes of 30 degree helix angle in $f = 0.33$ mm/flute, $d_a = 10$ mm, $d_r = 2$ mm and down cut.

With the cutter rotation angle ϕ measured from the bottom of the reference flute, the distance L_s from the clamped position of the cutter to the machining position of a flute is given as follow:

$$L_s = L_e - \frac{R}{\tan \alpha_h} \phi \tag{13}$$

where L_e is the effective length of the cutter. If the calculated cutting force is approximated by a uniform distributed force on the end mill corresponding to the axial depth of cut, the tool deflection induced at the machining position L_s is given by Eq. (14).

$$\delta_y = \frac{F_y/d_a}{24EI}[z_s^4 - 4L_e L_s^3 + 6L_e^2 L_s^2 - 4(L_e - d_a)^3 L_s + (L_e - d_a)^4] \tag{14}$$

Here, E and I are the elastic modulus and the moment of inertia of the cutter respectively. During rotation of the cutter, since the trace of tool deflection δ_y on the machined surface determines the surface profile, this equation provides the desired surface error. Some experiments were performed to verify the procedure as shown in Fig. 5, where the same cutting tool and conditions from the previous section were used. For measurements of surface errors, a digital probe having a resolution of 0.1 μm and diameter of 0.5 mm was used with computer interfaced data acquisition. The comparison between the predicted and the measured surface profiles shows nearly identical results in magnitude and shape. But in the cases of small axial depths of cut, some deviations in the surface profile were noticed. This was due to the fact that the approximation of uniform distributed cutting force

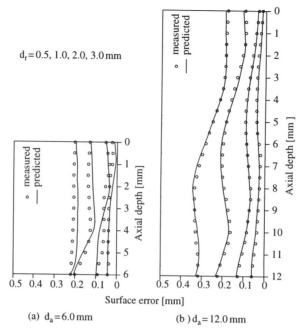

$d_r = 0.5, 1.0, 2.0, 3.0$ mm

(a) $d_a = 6.0$ mm   (b) $d_a = 12.0$ mm

Fig. 5.   Surface error prediction using the cutting depth estimation algorithm with an end mill of 6 mm diameter having four flutes of 30° helix angle and 30 mm effective tool length in 0.03 mm/rev feed, down cut and A12014-T6 work material.

may differ a little from the actual one for the relatively small axial depths of cut. The results reveal that the surface errors in end milling appear at arbitrary axial position.

This figure describes that though the waved shapes of these surface profiles change according to the variation of radial depth of cut, the surface profiles include each shifted component, as compared with the objective surfaces that would be produced by a completely rigid machining system.

2.3. *Feedrate adjustment*

It is possible to constrain the magnitude of surface errors within the specified tolerances by adjusting the feedrate along the tool path. The maximum error left on the workpiece surface is different at each location along the feed direction as a result of moving position of the force and the material removed from the workpiece. For a dimensional tolerance value on the workpiece surface, it is possible to schedule the feedrate along the feed axis in order to meet the required accuracy. Iteration or linear approximation algorithms can be applied for scheduling the feedrate.[6,23] In Fig. 6, the machining accuracies and the machining times are compared for the given workpiece with adjusted feedrates and conventional feedrates.

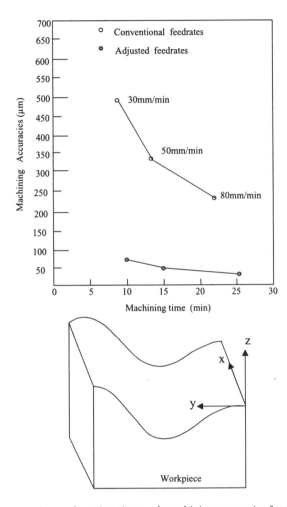

Fig. 6.   The comparisons of cutting time and machining accuracies for the workpiece.

3. Active Control Technique

3.1. *Tool deflection compensation*

The previous explanation implies that the surface error can be reduced by adjusting the relative positions of the tool with respect to the workpieces.

A small displacement at the end of the tool can be generated with a small angular displacement of the tool with respect to the center point, which is separated adequately from the end of the tool.

In order to implement this scheme in reducing machining errors through the adjustment of tool position, the tool deflection compensation system as shown in Fig. 7 was composed.

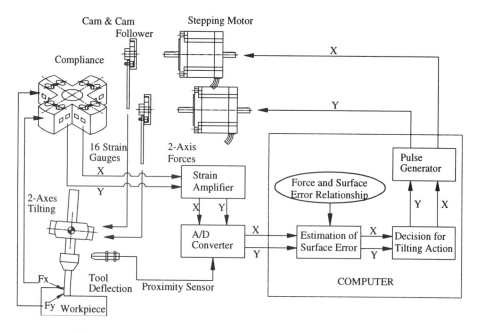

Fig. 7.   Configuration for the tool deflection compensation system.

This system is a computer integrated tool adapter including a force sensor and a tool tilting mechanism. The systematic procedure for the compensation of the tool deflection is as follows. The cutting forces generated in the end milling process are measured by the bi-axial force sensor enclosing the tool holder. The measured signals are then fed into the computer after a low pass filtering and A/D conversion. The angular positions of a flute on the end mill are monitored by a stationary proximity sensor located near the rotating trace of the protrusion on the tool holder. This information enables a point on the cutting force signal to synchronize with the angular position of a flute. With the aid of this information, the computer can now estimate the tool deflection and the machining error. The optimization procedure will be explained in the next section in detail. Using these results, it can control the tool position and thus minimize the expected surface error by sending out a tilting command through the interfacing board. This tilting action by the motor allows the tool to have a different radial depth of cut than before, thus a different tool deflection. The system then makes a loop, and again the process is repeated.

3.2. *Control algorithm*

As described in the preceding section, in machining, surface errors could be improved with the displaced tool positions corresponding to the errors. Based on this concept, a control plan for the compensation system was established, in which a control action is taken every couple of turns of the end mill. This method is expected to adapt to the variation of the cutting loads according to the shape of the workpiece and prevent the machining surface from being damaged with disturbances of the system due to the uneven cutting forces.

The block diagram of the control algorithm for the compensation system is shown in Fig. 8. The objective of this control flow is to achieve zero error value, $e(z)$, in the presence of tool deflection. Here, a digital PI controller as shown in Eq. (15) was implemented on the computer:

$$u(z) = \left(K_p + \frac{K_i}{1 - z^{-1}} \right) \cdot e(z) \tag{15}$$

where $u(z)$ is the controller output for driving the tilting actuator, $e(z)$ is the deviation indicating the surface error, and K_p and K_i are the proportional and integral gains, respectively. In the control flow, to estimate the deviation $e(z)$ from the reference of zero, a suitable value about the surface error is needed. As a suitable value, the minimum surface error, which is defined to be the minimum deviation from the surface to be produced without tool deflection, is selected. This is because the compensation relying on this value could be expected to be effective for obtaining a satisfactory machined surface without any overcut due to excessive compensation.

In this kind of control problem, i.e. geometrical adaptive control, in order to avoid the difficulty due to the direct sensing of the dimensional error, the estimation of the control parameter using a system model is frequently used. In the end milling, it is difficult to know the exact cutting size, like radial depth and axial depth of cuts, prior to the real cutting. Therefore the system predicts the control parameter, that is, the minimum surface error, based on the cutting force which is relatively easy to measure compared with measuring the error directly.

It is considered that the fluctuation of the cutting force becomes a correction factor in the estimation of the minimum surface error from the mean cutting force.

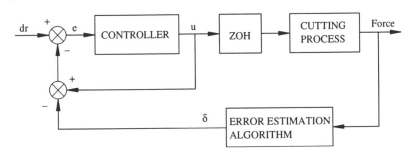

Fig. 8. System block diagram.

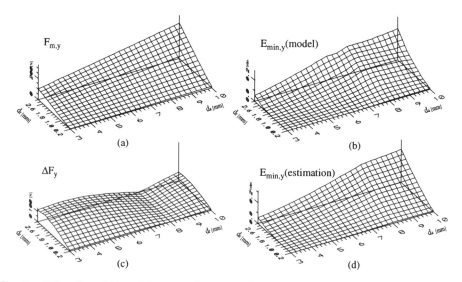

Fig. 9. Estimation of the minimum surface errors by cutting forces in a computer simulation. Applied cutting conditions are the same with those in Fig. 5.

Based on this, an empirical equation is suggested to evaluate the minimum surface error from the measured cutting force. This is shown in Eq. (16).

$$E_{min,y} = a \cdot F_{m,y} - b \cdot \Delta F_y \tag{16}$$

Here, the empirical constants a and b are affected by mechanical properties of the workpiece, the tool and the feedrate, but all of them are factors that can be fixed in the operating the system, and be valid in the change of cutting condition in radial depth and axial depth of cuts. Accordingly, if these are fixed, the empirical constants can be determined by the least square fitting method using the force and error data obtained from cutting experiments in ordinary cutting conditions. Figure 9(d) shows the result of the estimation by Eq. (16) and implies that this is an good approximation in comparing it with Fig. 9(b). It is also important to note that the linear form of this equation suggests a good feature, in which, even in the case that the normal direction of the machining surface does not coincide with the principal direction of the measured force, the 2-dimensional compensation using this equation is capable of being applied effectively to compensate the tool deflection in any lateral direction. From preceding experiments, the empirical constants a and b were determined as follows:

$$a = 1.05 \times 10^{-3}$$
$$b = 0.45 \times 10^{-3} \tag{17}$$

3.3. *Force sensor*

The cutting force can be measured by a force sensor.[24-26] The force sensor used to measure the lateral force acting on the tool was devised as an adapter built-in

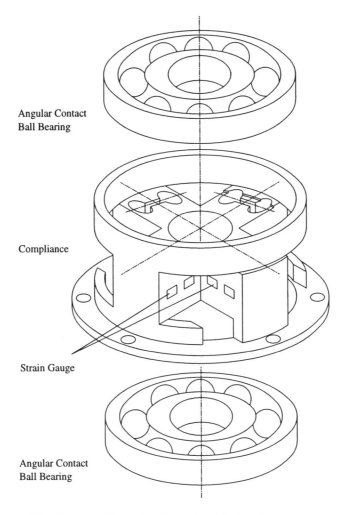

Angular Contact
Ball Bearing

Compliance

Strain Gauge

Angular Contact
Ball Bearing

Fig. 10.   Compliance for adapter built-in type force sensor.

type for compactness and easy application. The sensor was based on the principle of
the deformation of the parallel plate and the application of strain gauges since this
type has flexibility in design and is superior in long term stability. Figure 10 depicts
the compliance of the sensor, in which the basic structure is capable of detecting
bi-axial lateral forces without the coupling of the signal, which would be caused
by bendings, torsions and thermal deformations. With this design, the tool holder
is set up through the hole of the compliance, and the forces acting on the tool
are transferred to the compliance via the pre-loaded angular contact ball bearings.
Therefore, this structure allows the measurement of the forces on the rotating parts
at a stationary point. To measure the deformation of the compliance due to the
lateral cutting forces, a total of 16 strain gauges of which there are 4 to every 4

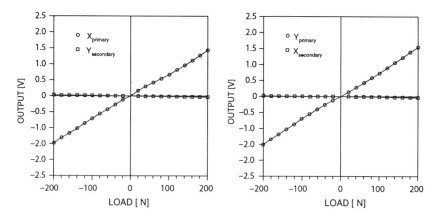

Fig. 11.   Static calibration of the devised sensor. (The plots for the unloading tests only slightly differ from the loaded test. To avoid confusion they have been omitted.)

wings of the compliance were applied with the Wheatstone bridge circuit, two to every arm.

The results of calibration for the completed sensor are shown in Fig. 11

The primary output of the sensor, which means the output from the principal sensible direction coinciding with the acting force in the design view point, has a linear relation with applied load. The secondary output, which is perpendicular to the direction of the acting force, were minimal. Also, to investigate the dynamic characteristics of the devised sensor, an impulse response from the sensor system including the compliance, the tool holder and a tool is measured in the circumferential direction of the tool holder with an acceleration pickup. As a result of the test, it is revealed that the compliance of the measurement system has its first natural frequencies of 574 Hz and 612 Hz in the two principal directions respectively. Considering a flute passing frequency of 20–80 Hz at an ordinary tool rotating speed of 300–1200 rpm, the dynamic characteristics of the devised sensor are expected to be sufficient for measuring the cutting force. Also, to examine its performance in real cutting, a test comparing directly it with a commercial sensor had been carried out. The commercial sensor used in this comparative tests was a 3-axes Kistler model 9257B piezoelectric dynamometer with a model 5007 charge amplifier. Figure 12 shows that there is no significant difference between the devised and the commercial sensor. Hence, a linear equation could be used for converting the sensor signal to the cutting force.

3.4. *Tool tilting device*

As a means for adjusting the tool position, the tilting method which makes an angular displacement to the tool instead of a direct translation was suggested. Here, a tilting mechanism to implement this idea was devised. Figure 13 represents the detailed drawing of the tilting mechanism. Referring to this figure, the tilting action

150
100
50
0
-50
FORCE [N]
Y devised
Y commercial
X devised
X commercial
1
1000
NO. OF SAMPLE

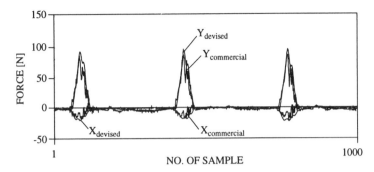

Fig. 12. Dynamic comparison of the devised sensor to a commercial dynamometer (f_s = 13.6 KHz, end mill of 2 flutes at 1200 rpm).

SCHEMATIC DIAGRAM
Tilting Cam
Roller Follower
Holder Shaft
Stepping Motor
Force Sensing
Compliance
Tilting Lever
Spherical Bearing
End mill
Spherical Joint
Guide Pin
Tool Holder
MOTION TRANSFER ROUTE
Stepping Motor -> Tilting Cam -> Roller Follower
-> Tilting Lever -> Spherical Bearing -> Compliance
-> Bearing -> Tool Holder -> Endmill

Fig. 13. Tool tilting mechanism.

of the tool is achieved as follows. The rotation of the stepping motor is converted into the linear motion of the tilting rod by the cam drive mechanism. At the same time, the driving force is amplified enough to operate against the cutting force, and also, the movement becomes more precise to obtain the required high resolution.

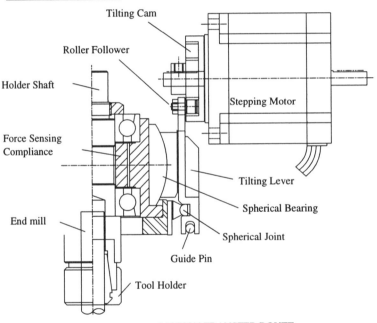

The linear displacement of the tilting lever is transformed into the angular displacement through the spherical bearing. Then the tool holder, which is located in the center of the spherical bearing and is connected with the spindle of the machine tools using a flexible coupling, obtains the tilting motion for the proper adjustment of the tool position. Since two sets of the tilting driver are actually located perpendicular to each other in the circumference of the spherical bearing, the angular motions as above in orthogonal direction make the system capable of adjusting the tool position in any direction. By the aid of this mechanism, in the case of using a tool having an effective length of 30 mm, the resolution for the one step (0.9 degree) of the stepping motor is 4 μm, and the range of moving to one side is 0.8 mm.

3.5. *Experiments*

In order to verify the ability of the system for suppressing the generation of surface errors by the tool deflection, experiments using this system have been carried out. Figure 14 shows a typical machining situation using the system.

In these experiments, the tool and equipment sets in the test of the previous sections were used. A feedrate of 72 mm/min and spindle speed of 600 rpm were applied. The control gains of K_p and K_i appearing in Eq. (15) were set to 0.15 and 1.0 respectively in the initial experiments. Three types of specimens that represent the shapes of workpieces commonly encountered in the end milling were prepared as

Fig. 14.   The computer integrated tool adapter and a typical machining using the system.

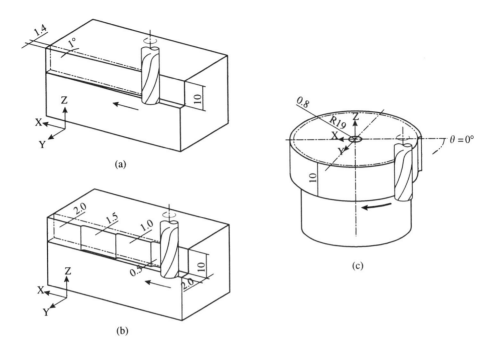

Fig. 15.   Specimens for tool deflection compensation tests.

shown in Fig. 15. In the figure, specimen (a), with a linearly increased radial depth of cut at a constant axial depth, was used to test the adaptability of the system for the gradual increments of the cutting load. Specimen (b), with a stepped radial depth of cut while maintaining an axial depth cut at a constant depth, was prepared to check the behavior of the system for the stationary and the abrupt changes of the cutting load. Specimen (c), which is a cylinder block, was subjected to a cutting load of a constant magnitude while its direction continuously changed. This specimen was selected to examine how well the system could handle the changes in all directions of the tool deflections.

Figures 16, 17 and 18 show the results of the experiments for specimens (a), (b) and (c) in Fig. 15, respectively. In these, the top figures represent the measured cutting forces and the system outputs for the compensation of the generated tool deflections. The middle figures in (a) and (b) depict the 3-dimensional error map of the measured errors respectively. The lower figures describe the measured minimum surface errors, in which the circle connected lines indicate the case where the compensation system was used and the triangle connected lines corresponds to the case where it was not used. These surface errors are the minimum values extracted from the measurement in the tool axial direction for the surface profiles at the reading points.

In Fig. 16, change of the cutting force due to the gradual increment of the cutting load lets the system predict the occurrences of the more machining errors and thus, the system sends larger compensating signals to the tilting actuator. In the case

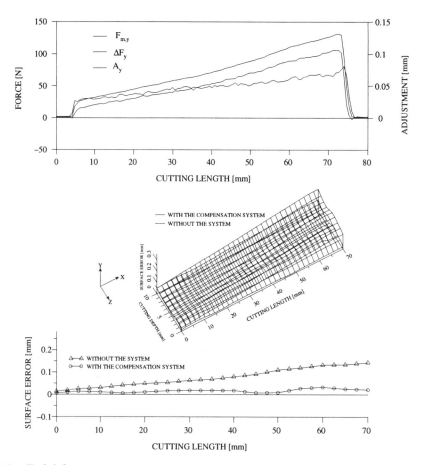

Fig. 16.   Tool deflection compensation tests and machined surface errors in the case of using vs. non using of the system for Specimen(a) in Fig. 15.

where the system was not used, the surface errors enlarge to 140 μm according to the increment of the radial depth of cut, while in the case where the system was used they are maintained within 30 μm. The results of the second test shown in Fig. 17 indicate that while the cutting loads have changed in stepwise, the mean cutting forces increase, but the fluctuating forces are almost level in the region where the radial depth of cuts are greater than 1.0 mm. On the other hand, there appear less machining error in the area where the cutting load are heavier. However, it should be pointed out that the value displayed in these figures is the minimum surface error in every measuring section, so the small value does not always mean a better surface accuracy. This can be shown in the 3-dimensional view found in the middle figures. But even here, it is also shown that the surface error are effectively suppressed by using the compensation system. In the result of the circular cutting test shown in Fig. 18, the cutting loads remain constant, while the locations of the machining surface with respect to the tool are continuously changed. In conforming

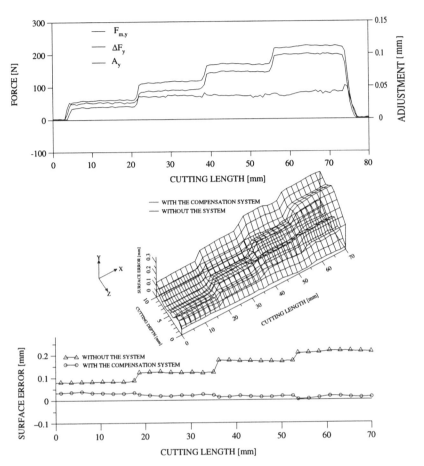

Fig. 17. Tool deflection compensation tests and machined surface errors in the case of using vs non using of the system for Specimen(b) in Fig. 15.

with that, the two directional cutting forces in each other orthogonal, which were measured relying upon a fixed coordinate system, show the curves of a sinusoidal form. The compensating command by the system also trace out the similar patterns with them. The measurement for the machined surfaces reveals that the errors over the circumference of the machined workpiece, in the case of not using the system, exhibit about 80 μm in the entire machined surfaces, while in the case of using the system are limited within 20 μm by the compensations of the tool deflections. This might appear frequently in the machining of plate cams, in these cases the suggested system can be applied successfully.

4. Conclusions

This chapter has presented two examples for reducing the machining surface errors due to the cutter deflection by the cutting forces in manufacturing environments.

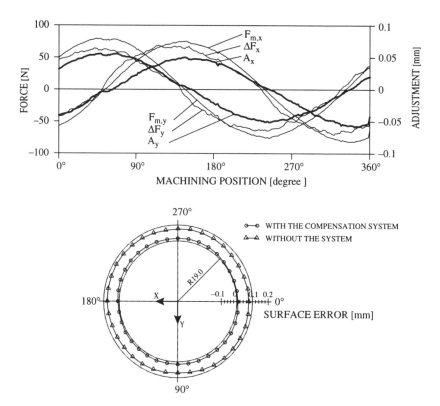

Fig. 18.   Tool deflection compensation tests and machined surface errors in the case of using vs non using of the system for Specimen(c) in Fig. 15.

The first example is the feedrate adjustment technique, in which the cutting forces are calculated and its machining accuracy is predicted by the been theory, and the feedrate is controlled by the required machining accuracy.

The second example is the active control technique, in which the system measure the cutting force, estimate the surface error and adjust the tool position on-line to reduce the machining error due to the tool deflection in the end milling process.

In the implementation of this task, the built-in type force sensor and the tilting device were devised along with some algorithms to operate the system.

The force sensor fixed on the housing of the tool adapter shows good performance in the measurements of the cutting forces activated on the rotating tools.

The active control system adapts properly to the cutting environments. The system proves to suppress the surface errors in the more productive cutting conditions without any alleviation of the cutting load.

References

1. Y. Altintas, Direct adaptive control of end milling process, *Int. Machine Tools Manufacturing* **34**, 4 (1994) 461–472 k.

2. Y. Altintas and I. Yellowley, The identification of radial width and axial depth of cut in peripheral milling, *Int. Machine Tools Manufacturing* **27**, 3 (1987) 367–381.

3. E. Budak and Y. Altintas, Peripheral milling conditions for improved dimensional accuracy, *Int. Machine Tools Manufacturing* **34**, 7 (1994) 907–918.

4. S. Y. Liang and S. A. Perry, In-process compensation for milling cutter runout via chip load manipulation, *ASME Journal of Engineering for Industry* **116** (May 1994) 153–160.

5. Y. S. Tarng and S. T. Cheng, Fuzzy control of feed rate on end milling operations, *Int. Machine Tools Manufacturing* **33**, 4 (1993) 643–650.

6. M. Y. Yang and C. G. Sim, Reduction of machining errors by adjustment of feedrates in the ball-end mill process, *Int. J. Prod.* **31**, 3 (1993) 665–689.

7. Z. Yazar, K. F. Koch, T. Merrick and T. Altan, Feed rate optimization based on cutting force calculations in 3-axis milling of dies and molds with sculptured surfaces, *Int. Machine Tools Manufacturing* **34**, 3 (1994) 365–377.

8. D. A. Milner, Controller system design for feedrate control by deflection sensing of a machining process, *Int. J. Mach. Tool Des. Res.* **14** (1974) 187–197

9. M. A. Elbestawi, Y. Mohamed and L. Liu, Application of some parameter adaptive control algorithms in machining, *ASME Journal of Dynamic System, Measurement, and Control* **112** (December 1990) 611–617.

10. T. Watanabe and S. Iwai, A control system to improve the accuracy of finished surfaces in milling, *ASME Journal of Dynamic Systems, Measurement, and Control* **105** (September 1983) 192–199.

11. M. Y. Yang and J. G. Choi, A tool deflection compensation system for end milling accuracy improvement, *ASME Journal of Manufacturing Science and Engineering* **120** (May 1998) 222–229.

12. E. J. A. Armarego and N. P. Deshpande, Computerized end-milling forces predictions with cutting models allowing for eccentricity and cutter deflections, *Annals of the CIRP* **40**, 1 (1991) 25–29.

13. A. Ber, J. Rotberg and S. Zombach, A method for cutting force evaluation of end mills, *Annals of the CIRP* **37**, 1 (1988) 37–40.

14. W. A. Kline, R. E. Devor and J. R. Lindberg, The prediction of cutting forces in end milling with application to cornering cuts, *Int. J. Mach. Tool Des. Res.* **22**, 1 (1982) 7–22.

15. F. M. Kolarits and W. R. DeVries, A mechanistic dynamic model of end milling for process controller simulation, *ASME Journal of Engineering for Industry* **113** (May 1991) 176–183.

16. C. G. Sim and M. Y. Yang, The prediction of the cutting force in ball-end milling with a flexible cutter, *Int. Mach. Tools Manufact.* **33**, 2 (1995) 267–284.

17. S. Smith and J. Tlusty, An overview of modeling and simulation of the milling process, *ASME Journal of Engineering for Industry* **113** (February 1991) 169–175.

18. M. Y. Yang and H. D. Park, The prediction of cutting force in ball-end milling, *Int. Mach. Tools Manufact.* **31**, 1 (1991) 45–54.

19. W. A. Kline, R. E. Devor and I. A. Shareef, The prediction of surface accuracy in end milling, *ASME Journal of Engineering for Industry* **104** (August 1982) 272–278.

20. T. Matsubara, H. Yamamoto and H. Mizumoto, Study on accuracy in end mill operations (1st Report) — Stiffness of end mill and machining accuracy in side cutting, *Bull. Japan Soc. of Prec. Engg.* **21**, 2 (June 1987) 95–100.

21. T. Matsubara, H. Yamamoto and H. Mizumoto, Study on accuracy in end mill operations (2nd Report) — Stiffness of end mill and machining accuracy in side cutting, *Bull. Japan Soc. of Prec. Engg.* **25**, 4 (December 1987) 291–296.

22. J. W. Sutherland and R. E. Devor, An improved method for cutting force and surface error prediction in flexible end milling systems, *ASME Journal of Engineering for Industry* **108** (November 1986) 269–279.

23. E. Budak and Y. Aetintas, Modeling and avoidance of static from errors in peripheral milling of plates, *Int. J. March. Tools Manufact.* **35**, 3 (1995) 459–476.

24. S. E. Oraby and D. R. Hayhurst, High-capacity compact three-component cutting force dynamometer, *Int. Mach. Tools Manufact.* **30**, 4 (1990) 549–559.

25. Y. Tani, Y. Hatamura, and T. Nagao, Development of small three-component dynamometer for cutting force measurement, *Bulletin of the JSME* **26**, 214 (April 1983) 650–658.

26. J. H. Tarn and M. Tomizuka, On-line monitoring of tool and cutting conditions in milling, *ASME Journal of Engineering for Industry* **111** (August 1989) 207–212.

CHAPTER 7

TECHNIQUES AND APPLICATIONS OF ON-LINE PROCESS QUALITY IMPROVEMENT

GANG CHEN

Gensym Corporation,
125 Cambridge Park Drive,
Cambridge, MA 02140, USA
Tel: (617) 588-9457; Fax: (617) 547-1962
gchen@gensym.com

Producing good quality products is an important process control objective. However, achieving this objective can be very difficult in a continuous process, especially when quality measurements are not available on-line or they have long time delays. At the same time, process safety is also a critical issue. It is important to monitor process performance in real time. Here, a real time process quality monitoring and a control approach using multivariate statistical models are presented to achieve this objective. The goal of the monitoring approach is to detect faults in advance and the control approach is to decrease variations in product quality without real time quality measurements. A principal component analysis (PCA) type of model which incorporates time lagged variables is used, and the control objective is expressed in the score space of this PCA model. A process quality monitor is developed based on the process description of this PCA model. A controller is designed in the model predictive control (MPC) framework, and it is used to control the equivalent score space representation of the process. The score predictive score model for the MPC algorithm is built using partial least squares (PLS). The proposed controller can be developed from and implemented on top of existing PID control systems. The proposed monitor and controller are demonstrated in case studies, which involve a binary distillation column and the Tennessee Eastman process.

Keywords: Multivariate statistics; PCA; multi-way PCA; PLS; MPC process control.

1. Introduction

On-line monitoring of chemical process performance is extremely important for plant safety and good product quality. In a typical chemical plant, process computers collect many observations very frequently on process variables such as temperatures, flows, and pressures, etc. However, most quality variables, which are the key indicators of process performance, are the result of sample analyses made off-line in a quality control laboratory. The use of such sampled inputs as performance index can delay the signaling of faulting situations. Things could be even worse when such data are used as feedback signals for conventional control systems. It can cause major and prolonged deviations from their set points, since disturbance effects remain undetected in between the sampling times. When analytical sensors are unavailable, a mathematical representation of a process can be used for describing the process performance. The representation of a process can take two forms, a first principle model, or an empirical model based on historical data. Many factors, such as high process nonlinearity, high dimensionality, and the complexity of a process, can make the development of a first principle model very difficult. As an alternative, empirical modeling approaches that are basically data-driven multivariate statistical methods have attracted much interest by chemical engineers.[13,15,18,20–22] These approaches are based on the theory of Statistical Process Control (SPC), under which the behavior of a process is modeled using data obtained when the process is operating well and in a state of control. Future unusual events are detected by referencing the measured process behavior against this model. These approaches use PCA type method to model the process. Because of the nice features of PCA, their methods can handle high dimensional and correlated process variables. The key point of the approach is to use PCA to compress normal process data and extract information by projecting the data onto a low dimensional score space. New multivariate SPC charts, which are similar to conventional Shewart charts,[24] have been developed for monitoring in the low dimensional space. Simulation results show that multivariate methods are simple and powerful. Nomikos and MacGregor[20] present a Multi-way Principal Component Analysis (MPCA) approach for monitoring batch processes. The reference data set for a batch process inherently is time varying in nature, and it can be arranged as a three-dimensional array. The three dimensions involve the process variables, time through a batch, and batch number. MPCA can handle a three-dimensional array of data and it is statistically and algorithmically consistent with PCA. Furthermore, an on-line monitoring strategy that has characteristics similar to the dynamic matrix control (DMC) algorithm is proposed for batch monitoring by Nomikos and MacGregor.[20] They predict the behavior of a batch at its end point, and this predictive feature of their strategy is promising for use in continuous processes. Here, the monitoring method using an MPCA model is extended to continuous processes. The model uses time lagged data, and so the model inputs are dynamic. Thus, this method overcomes the steady state assumption often used in PCA and it can be used for real time monitoring. No data filtering is needed, and a predictive monitoring strategy is proposed. The monitoring goal is to rapidly detect important process shifts and the occurrence of new disturbances. This modeling technique can also be used in quality control problem. A new secondary variable approach is studied for use when quality measurements are not available on-line or have long time delays, and a multivariate statistical controller is proposed. The key aspects of this controller are as follows. A MPCA model that uses time lagged data is used[16] and the scores calculated from this model are fed as inputs to a score predictive model which is developed using partial least squares (PLS). The predicted scores are used as key indicators of the process performance based on the assumption of an implicit correlation between available measurements and quality variables. The control objective is defined as maintaining the predicted score variables within a certain acceptable region defined from historical data.[23] This controller is developed from and implemented on top of an existing, plant-wide conventional PID control system. Manipulated variables for the proposed controller are selected set points of existing control loops. Experimental testing is only needed on these selected set points. To drive the predicted scores close to their set point, model predictive control (MPC) is used as the control algorithm. Other issues, such as how to retain the correlation structure of the input variables when implementing the controller and how to eliminate steady state offset of the quality variables, are also addressed. In the following sections, the techniques used in this study are briefly reviewed, and the design of the multivariate statistical monitor and controller is discussed in detail. A distillation column and the Tennessee Eastman process[5] are used to demonstrate the effectiveness of the controller. The final section draws conclusions regarding this study.

2. Multivariate Statistical Techniques

In this section, a brief overview of the multivariate statistical techniques used, namely PCA and PLS, is presented. These techniques can be used to transform noisy and correlated measurements into a smaller informative set. This transformation is achieved by identifying the underlying phenomena within the actual measurements in a reduced dimensional subspace. Principal component analysis (PCA) was originally developed by Pearson (1901). The details of linear PCA can be found in Jolliffe[12] and Jackson.[11] The development of PLS is largely due to the work of Wold (1966). A tutorial description of the PLS algorithm is given by Geladi and Kowalski.[9]

2.1. *Principal component analysis*

Assume that process data are given in a two-dimensional data matrix $X(m \times n)$, where m stands for the number of samples, and n stands for the number of variables. PCA is a method used to explain the variance of X in terms of new latent variables called principal components. If there is correlation, then a small number of principal components will summarise a majority of the variation in X. Changes that occur within the principal components can be used to analyse changes in the original data space. Computationally, PCA is usually handled by computing eigenvectors of the covariance matrix of X or by the singular value decomposition (SVD) algorithm. Principal components, also called score variables, are the projections of original variables along the directions determined by the first k largest eigenvectors $(p_1 \ldots p_k)$ of the covariance matrix of X. The k principal components define the subspace with the greatest variability among all possible k dimensional subspaces projected from X. The eigenvectors $(p_1 \ldots p_k)$, also called the loading vectors, give the directions of the subspace, and each observation is located in this subspace via its scores. Because of nonlinearity in real processes, it may be necessary to use nonlinear models. The principal curve[10] is a generalisation of a linear principal component, but when applied to data sets the principal curve algorithm does not yield a nonlinear PCA model, but rather a table of results. Dong and McAvoy[4] proposed a nonlinear PCA (NLPCA) method which integrates the principal curve algorithm and neural networks. NLPCA models are built by using neural networks after the principal curve values are obtained.

Ordinary PCA is good at analysing a two-dimensional data matrix $X(n \times m)$. In many cases, especially in areas like chemistry, psychometrics, and image analysis, data from experimental studies are three-dimensional. Multivariate time series analysis is one of the problem areas where the data can be structured in three dimensional array, as show in Fig. 1. PCA cannot be applied directly to such a three-dimensional data array. A modified PCA algorithm, namely Multi-way PCA (MPCA), gives an approach to summarise the variation within a three-dimensional array and to represent one sample of X by a single score vector. The MPCA algorithm for computing principal components for three-dimensional arrays is described in Wold et al.[27] It has been proven that MPCA is equivalent to performing PCA on a large two dimensional matrix formed by unfolding the three-dimensional array.[8,25,27] The idea behind this is quite simple: define any two of the three dimensions so that they form one single dimension. Figure 2 shows one method of unfolding a data array. After unfolding, either linear PCA or nonlinear PCA[4] can be performed as the MPCA calculation.

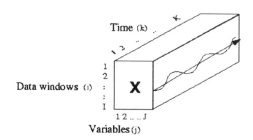

Fig. 1.   A three dimensional array.

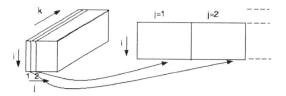

Fig. 2.   Array unfolding.

2.2. *Partial least squares*

In many cases, one can identify two sets of data: an input data set X, and an output data set Y. The problem is not just to explain the variability in X, but to use X to predict or infer the output Y. Partial least squares (PLS) is a method for modeling the relationship between these two sets of data. Conceptually PLS is similar to PCA except that PLS simultaneously reduces the dimensions of the X and Y spaces to find latent vectors for the X space which are predictive of the Y space. The PLS algorithm is based on decomposing the X and Y matrices into two low dimensional score matrices, T and U, and loading matrices, P and Q, with the constraint that the score vectors of the same component are related. This connection is called the inner relation of the X and Y matrices. The mathematical expressions are as follows:

$$X = TP^T + E$$
$$Y = UQ^T + F$$

where E and F are residuals. The inner relation is given by:

$$U = TB + H$$

where H is the residual. This linear regression forces the score vectors u_i and t_i in U and T to be collinear, and thereby maximises the covariance. This technique works well when dealing with a small number of observations that are highly correlated and have noise in both X and Y. To handle process nonlinearity, Qin and McAvoy (1992) proposed a neural net PLS (NNPLS) modeling approach, which is an example of integrating neural networks and statistical techniques. Nonlinearity is incorporated by using neural networks to model nonlinear inner relationships. A theoretical analysis and applications of NNPLS can be found in Qin and McAvoy (1992).

2.3. *Model predictive control*

In this study, a multivariate statistical controller is designed in the model predictive control (MPC) framework. Many MPC algorithms using linear or nonlinear models have been proposed. Using a linear step response model, dynamic matrix control (DMC)[3] is one of the earliest MPC algorithms. DMC embodies many concepts that are present in other predictive control algorithms. The original DMC algorithm formulates an unconstrained control problem for multi-input-multi-output (MIMO) systems. In 1986, Garcia and Morshedi modified the DMC algorithm so that constraints on the plant input and output variables are explicitly included in the problem formulation. The resulting optimisation problem is solved by quadratic programming (QP) techniques, and this control method is referred to as QDMC. While MPC algorithms using linear models (denoted as LMPC) have significant advantages for implementation as well as theoretical analysis, algorithms using nonlinear models (denoted as NLMPC) may be necessary in practice since many chemical processes are inherently nonlinear. NLMPC algorithms can be formulated using nonlinear programming (NLP) techniques, in which an objective function is minimized to calculate the future control moves. Notice that QDMC (although using a linear model) also fits in this structure where the optimizer is a QP solver. A theoretical analysis and controller design for MPC can be found in Garcia *et al.*[7]

3. Continuous Process Monitoring Strategies

Because of the wide use of process computers, data based statistical approaches have great potential for chemical process monitoring. A PCA monitoring method proposed by Kresta *et al.*[14] is an excellent example of these approaches. The basic methodology applied in Kresta *et al.*'s approach is similar to the Shewart chart method. A Shewart chart consists of plotting variable observations sequentially on a graph, which also contains the target value and upper and lower control limits. If the observations exceed the control limits, some abnormal operation is assumed to have occurred. The control limits are usually determined by analysing a reference set of process data collected when the process is operating well and in a state of control. However, this univariate approach is not satisfactory for chemical processes because of the high dimensionality of process variables. Kresta *et al.*'s approach uses PCA to compress the normal process data and extract information by projecting the data onto a low dimensional score space. Two types of SPC charts are developed for continuous process monitoring. One is a score plot, and the other is a Square Prediction Error (SPE) plot. The axes of a score plot are formed by score variables and each observation is located on this plot via its scores. The SPE is given by:

$$SPE = \sum_{i=1}^{m}(x_i - x_i')^2$$

where x_i, $i = 1, \ldots, m$ is a PCA input, and x_i' is the prediction of x_i from the PCA model. The SPE can be plotted against time to form an SPE plot. The control limits in the score plot and the SPE plot are determined based on the reference distribution for the data set determined from normal operation. An important issue in using Kresta *et al.*'s[14] approach for continuous process monitoring is the development of a reference data set. Because ordinary PCA analyses the variance of each individual sample separately, the reference data used as the inputs to a PCA model are usually steady state data. One assumption of Kresta *et al.*'s approach is that the process under investigation is at steady state. Extracting steady state information from normal operating data requires data filtering. One filtering technique, namely the FIR Median Filter Hybrid method, was used by Piovoso *et al.*[21] to develop a PCA model for continuous process monitoring. To overcome the steady state assumption and analyse dynamic data for continuous process monitoring one possible method is to use time lagged data.[16]

Because the data from continuous processes are usually time varying, MPCA's ability to analyse the time histories of the variables is very useful for continuous process monitoring. In dynamic systems, the time dependency is seen in the variable histories. To monitor a dynamic process, not only is one concerned with the correlation among variables, but also with the auto correlation of each variable. As a result, a dynamic model is needed for the monitoring task. In the theory of system identification, Leontaritis and Billings[17] have proposed an Auto-Regressive, Moving Average, eXogenous (ARMAX) model structure, which can represent a wide class of dynamic systems. The inputs, $x(t)$, for an ARMAX model can be described mathematically in the following general form:

$$x(t) = [y_m(t), \ldots, y_m(t - q)]$$

where y_m is the vector of measurements. The parameter q determines the *order* of the ARMAX model, and t stands for the current time. $x(t)$ contains samples of y_m from time $t - q$ to t, and $x(t)$ can be defined as a data window in the time dimension, as shown in Fig. 3. With an appropriate order, which is the window length, a data set with this structure can capture the process dynamics. The data windows are obtained by cutting along the time dimension of the data series. The reference set is developed by stacking the windows together. The procedure is shown in Fig. 4. The data windows should overlap in order to produce smooth predictions. That is, one data window should begin before the proceeding one ends. This point is discussed in the example considered below. The reference set is a three-dimensional array,

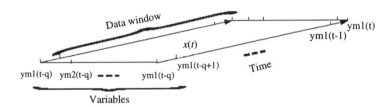

Fig. 3.   A data window for a continuous process.

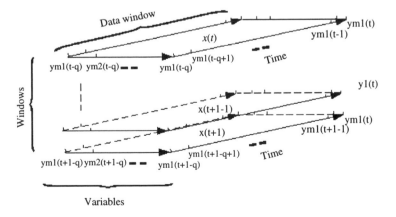

Fig. 4.   The development of a reference set.

which is the same as that shown in Fig. 1, where J is the number of variables, K is the time interval through a data window, and I is the number of windows.

For the purposes of dynamic process monitoring, the reference set is unfolded in the same way as shown in Fig. 2. After unfolding, either linear PCA or nonlinear PCA[4] can be carried out as the MPCA calculation. The principal component variables explain the maximum variance of X over the time horizon of data windows. Changes that occur within the principal components are used to monitor the changes in the original data space. Similar to Kresta *et al.*'s[14] approach, the statistics used for the monitoring task are the score variables and the value of the Squared Prediction Error (SPE). The SPE is given by:

$$SPE = \sum_{i=1}^{m}(x_i - x_i')^2$$

where x_i, $i = 1, \ldots, m$ is an MPCA input, and x_i' is the prediction of x_i from the MPCA model. The distribution for the scores can be well approximated by a normal distribution, and the SPE can be viewed as having an $\alpha\chi^2$ distribution, whose parameters can be estimated from historical data. The control limits for the score variables and the SPE can be calculated from their approximate distributions. In using score and SPE plots, a reference set defines a normal operation range, and new process data are monitored against this normal range to detect abnormalities. In general, there are two kinds of abnormalities in a process. For the first kind of abnormality, the basic relationship among the variables and their dynamic patterns in a window do not change, but several variables have a larger than normal change. For this kind of abnormality, the SPE remains small, but the principal scores, t_i, move outside the region over which the model was developed. The second kind of abnormality results when the relationship among the variables or their dynamic patterns in a window change. For this kind of abnormality, the SPE increases, since the new correlation among the measurements cannot be explained by the MPCA model. When an MPCA model for continuous processes is developed, the variables should be scaled relative to one another, so as to avoid having important variables whose magnitudes are small from being overshadowed by less important but larger magnitude variables. The variables are scaled to zero mean and unit variance. The scaling of continuous process data is carried out with the assumption that the mean and variance of each variable remain the same at every time interval through every window.

For continuous processes, the data window can be moved forward sample by sample. At every new sampling time, there is one data window with its latest point at the present sampling time. For this window, a score point and an SPE can be calculated. If the dynamic pattern within the window is different from that of normal operation, or some variables have larger than normal deviations from their means, the SPE value for this window will increase or its score point will move outside the normal region over which the model was developed. These observations form the basis for a real time monitoring strategy. This real time monitoring method can handle dynamic data and as discussed later on it has a timing advantage over traditional monitoring approaches, which do variance analysis on individual data points using PCA models.

3.1. *Predictive monitoring approach*

Quick detection of process faults is a very important requirement of dynamic monitoring. A predictive monitoring strategy for continuous processes is proposed here. In this predictive monitoring approach, a data window containing predicted future data is projected onto the low dimensional score space developed from the reference set. The length of the data window is q and the predictive horizon within this window is h, where $h < q$. A score and an SPE can be calculated for this window corresponding to time $t + h$, where t is the current time. In this way, predictive monitoring is carried out. To have such a window, one must replace h future observations with appropriate values such that the predicted score for time $t + h$ will be as close as possible to that which would be calculated if the data up to time $t + h$ were known. Many predictive schemes, such as linear extrapolation, a Kalman filter, and a neural network model, can be used to predict the future observations. Here a simple predictive approach used by Nomikos and MacGregor[20] [shown in Fig. 5(a)] is proposed for continuous processes. The assumption is that the future deviations of the MPCA input variables from their means remain constant at their current values for the rest of the window. Figure 5(b) illustrates this predictive strategy for continuous processes. By contrast to a batch process, the prediction remains constant since the mean of each variable is assumed to be constant with time. Although this method is very simple, the assumption involved may produce data patterns that are not included in the reference set. Thus, the predictions may only be valid for a portion of a data window. If too many data points in a data window are predicted, the scores or the SPE for this window will be out of their control limits even when there is no fault in the available data. A maximum value

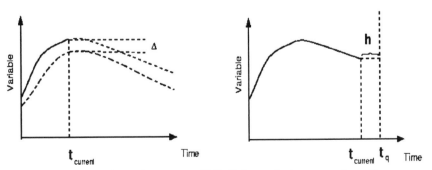

(a) On-line monitoring strategy for batch processes   (b) Predictive monitoring strategy for continuous processes

Fig. 5.   Monitoring strategies.

of h should be determined when the MPCA model is developed. How to select this maximum value is discussed in the case study.

3.2. *Model development and implementation procedure*

The procedure for continuous process monitoring will be illustrated using a case study. A simple description of this procedure is given as follows.

- The development of normal operating data. Kresta *et al.*[14] point out that for a data based method, the model developed will only be applicable to the conditions under which the data are gathered. The normal data should be selected for conditions which encompass low and high through puts of process variables and which yield desired product quality outputs.
- The data pretreatment. Data pretreatment is an important issue because the technique used here is data driven. For continuous processes, scaling is done before the unfolding process, since each variable has same statistical mean and variance for every window. Because the model has a dynamic structure, there is no need to extract steady state values from data sequences by using filtering techniques. However, outlier elimination is an important issue that should be addressed. When an MPCA model is developed, unrealistic values should be eliminated from the reference set.
- The development of the MPCA model.
- The development of score and SPE plots.
- Determining the maximum value of the predictive horizon $h_{\max}$. A new test set of normal data is used for determining the maximum h. A data window, which is established from the normal data, has h predicted samples using the algorithm discussed above. If h is small, the scores and the SPE corresponding to this data window are within their control limits. When h is larger than a certain value $h_{\max}$, the scores or the SPE corresponding to this data window will be outside their control limits even when this data window is based on a normal data set. The reason for this violation is that by predicting too many future values one generates records that differ from normal plant operation. The value $h_{\max}$ is defined as the maximum value of the predictive horizon, which does not produce false alarms.
- The implementation of the monitoring approach. The real time implementation of the monitoring approach is as follows. For each new sample, a window with previous data and new data is established. The score and the SPE for this window are calculated and then real time monitoring is carried out. The resulting data window gives the monitoring result at the current time and this result is usually very accurate. Then a data window with h predicted samples is established for predictive monitoring. h is equal to or less than the maximum predictive horizon $h_{\max}$. Predictive monitoring provides an early detection of process faults. However, when h is near $h_{\max}$, the predictive monitoring approach may produce some false alarms.

The proposed monitoring approach is illustrated by a realistic example below.

3.3. *Example: The Tennessee Eastman Process*

The Tennessee Eastman Process involves a simulation of a real plant that has been disguised for proprietary reasons.[5] The process produces two products, G and H, from four reactants, A, C, D, and E. Also present are an inert, B, and a by-product, F. The process has five major units: a reactor, a product condenser, a vapor/liquid separator, a recycle compressor, and a product stripper. The gaseous reactants are fed to the reactor where they react to form liquid products. The gas phase reactions are catalysed by a non-volatile catalyst dissolved in the liquid phase and the catalyst remains in the reactor. The products leave the reactor as vapors along with unreacted feeds. The process has 41 measurements and 12 manipulated variables. The modes of process operation are set by the G/H mass ratios. The product mix is normally dictated by product demands. The plant production rate is set by market demand or capacity limitations. There are six modes of process operation at three different G/H mass ratios (stream 11) as explained by Downs and Vogel.[5] There are two major product quality variables in the process, product flow rate (Stream 11) and product G/H ratio. McAvoy and Ye[28] presented a base control system, shown in Fig. 6, which is able to reject all important process upsets and meet all of the specifications placed on the problem. In this study, a 50/50 G/H product ratio under the base control system is considered. Normal operation

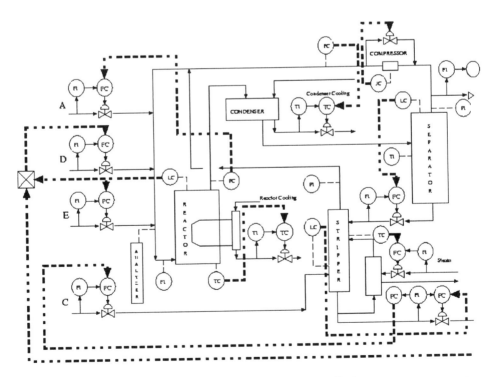

Fig. 6.   The base control system for the Tennessee Eastman process.

Table 1.   Inputs of the MPCA model.

Process variables	
XMEAS(1)	A Feed (stream 1)
XMEAS(2)	D Feed (stream 2)
XMEAS(3)	E Feed (stream 3)
XMEAS(4)	A and C Feed (stream 4)
XMEAS(5)	Recycle Flow (stream 8)
XMEAS(6)	Reactor Feed Rate (stream 6)
XMEAS(9)	Reactor Temperature
XMEAS(10)	Purge Rate (stream 9)
XMEAS(11)	Product Sep Temp
XMEAS(13)	Prod Sep Pressure
XMEAS(14)	Prod Sep Underflow (stream 10)
XMEAS(16)	Stripper Pressure
XMEAS(18)	Stripper Temperature
XMEAS(19)	Stripper Steam Flow
XMEAS(21)	Reactor Cooling Water Outlet Temp
XMEAS(22)	Separator Cooling Water Outlet Temp

is defined as the control response to random A, B, C composition upsets in the feed stream 4 (disturbance IDV(8)). A total of 16 input variables are available in the Tennessee Eastman process for building the MPCA model. These 16 input variables include the two used for manipulating the G/H ratio, variables not used in the base control system, as well as controlled variables of inner cascade loops, and they are listed in Table 1. Controlled variables, such as reactor level and reactor pressure, are strictly under control, so they are not suitable for detecting faults.

The reference set contains 1000 samples from normal operation with a sampling interval of 5 min. Since the time constants of the control loops range from several minutes to more than 10 h, a window length of 100 min (20 steps) is selected. Every two joined data windows have 18 steps of overlap. Even though each new window has only 2 new time values, one cannot get 500 windows from 1000 samples since information is not available at the end of the data set. Thus the reference set X is an (491 × 16 × 20) array, where 491 is the number of available data windows, 16 is the number of variables, and 20 is the window length. In this example, 18 steps of overlap are used for modeling. In using the resulting model one new set of data is used at each time step. Thus, the use of data in the modeling approach is close to how the data is used during real time operation. In general, it is probably good to use significant overlap in developing an MPCA model since the overlap should increase the richness of the training data set.

A linear MPCA model is developed from the (491 × 16 × 20) data array. Five principal components are selected, which capture 67.47% of the variation in the reference set. The control limits shown in every plot correspond approximately to the 95% and 99% confidence regions based on a normal reference distribution. These control limits are determined by using the methodology presented by Nomikos and MacGregor.[20] The plots of the first two normal scores $(t1, t2)$ and the SPE versus

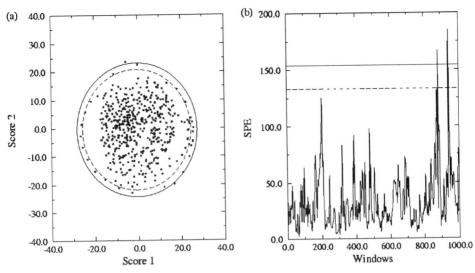

Fig. 7.   The score plot and SPE plot of normal operating data, (a) the score plot $(t1, t2)$, (b) the SPE plot. (Dash line: 95% control limit, solid line: 99% control limit.)

time are shown in Fig. 7. The dash line stands for the 95% control limit, and the solid line stands for 99% control limit. In all the figures, the control limits are drawn the same as in Fig. 7.

First, the maximum predictive horizon $h_{\max}$ is determined. A new set of "normal" data with 50 data windows, corresponding to 590 minutes, is tested against the reference set. The evaluation of the "normal" data is carried out with the predictive horizon h ranging from 0 to 15 steps. When the corresponding scores and SPEs are placed in the score and SPE plots, which are not shown in this paper, one can see that 95% of the scores remain within the region of their 95% control limit, and 95% of the SPEs lie below their 95% control limit with the predictive horizon less than or equal to 10 steps. If the predictive horizon is larger than 10, the SPE from that window will be above its 95% control limit, and in some cases the corresponding score may also be out of its 95% control limit. The test is carried out at 10 different sampling times, and the results are essentially the same. These results indicate that the maximum predictive horizon for the proposed monitoring strategy is 10 steps (50 min) and that using a predictive horizon of 10 steps or less will result in a new set of good data lying within the acceptable range of the variation defined by the reference set.

Three abnormal cases are tested for evaluating the ability of the dynamic monitoring approach to discriminate between "normal" and "abnormal" dynamic behavior. In the first case, the disturbance IDV(8) is combined with a large step change in A/C composition ratio in feed stream 4. This step is three times larger than that of the original disturbance IDV(1) (disturbances IDV(1) $\times$ 3 + IDV(8)). The second case involves a large step change in the B composition in feed stream 4, whose magnitude is three times larger than that of the original IDV(2), and the disturbance IDV(8) (disturbances IDV(2) × 3 + IDV(8)). The original magnitudes of the disturbances IDV(1) and IDV(2), which are about 10% for the A/C ratio and about 100% for the B composition, are defined in the problem description.[5] In the third case, the analyser for B in the purge flow is assumed to have a drift. The readings of the B composition drift away from their true value at a rate of 10% of the initial steady state value in 10 h. In the simulation, all three abnormal cases start after 100 min of operation (at the 20th sampling point). In the first two cases, the process goes to a region different from that of normal operation, but it is still under control. The drift of the B composition measurement in the purge is a serious disturbance. It will cause the process to shutdown due to a pressure build up unless the operator makes a change in operation.

Above, it is pointed out that there are generally two kinds of abnormalities in a process. For the first kind of abnormality, the basic relationship among the variables and their auto correlation within a window do not change, but several variables have a larger than normal deviation from their means. In this case the SPE remains small, but the scores move outside the normal operating region. The second type of abnormality results when the relationship among the variables or their dynamic patterns in a window change. For this kind of abnormality, the SPE increases. The first test of these three abnormal cases uses the real time monitoring strategy, in which the windows contain all known data up to current point, i.e. there is no prediction. This approach is identical to the dynamic PCA approach of Ku et al.[16] Figure 8 shows the results of case 1, and similar results are achieved for cases 2 and 3. The SPE values in all cases exceed the 99% control limits after about 250 to 350 min (50–70 steps), which indicates that the dynamic behavior of all three cases is not consistent with that in the reference set. In Fig. 8, the score is outside of its 99% control limit at step 65 and the SPE is above its 99% control limit at the 88th step. These results indicate that a large IDV(1) will cause more deviation of the measurements from their means than changes of the dynamic patterns in the measurements.

To verify that the dynamic monitoring approach using MPCA is useful for detecting faults, the monitoring strategy discussed above is compared with the PCA monitoring method presented by Kresta et al.[14] for continuous processes. To extract steady state values from the reference data set, a discrete filter — FIR Median Filter Hybrid method[21] is used. Real time monitoring is carried out on the current sample vector. Figure 9 gives the results for case 1. The PCA model used has three principal components, which explain 71.2% variation in the reference set. Comparing the SPE plot in Fig. 9 to that in Fig. 8, one can see that the real time monitoring approach using MPCA detects the faults earlier than the steady state approach. Similar results are achieved for cases 2 and 3. The same conclusion can be drawn from the score plots. The advantage of the dynamic MPCA method over the standard PCA method results from using the joint covariance matrix of all the variables and their histories. The dynamic MPCA method utilises not only the

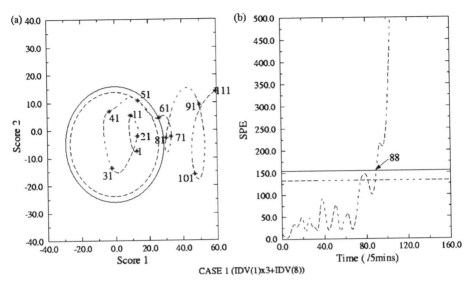

Fig. 8.   The result of real time monitoring for case 1, (a) score plot, (b) SPE plot. (The control limits are the same as in Fig. 7.)

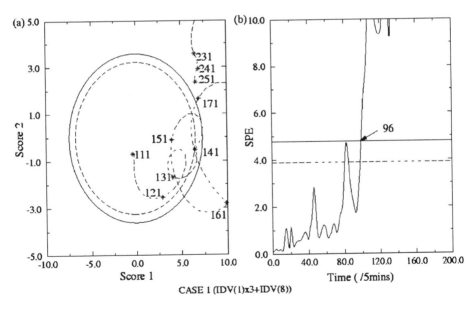

Fig. 9.   The result of PCA time monitoring for case 1, (a) score plot, (b) SPE plot.

relationships among variables but also the auto correlations within each variable in order to detect abnormal operation.

The timing advantage is even better when the predictive monitoring strategy is applied to the three cases. The result with a predictive horizon of 10 steps is shown in the SPE plots in Fig. 10. In this plot, the solid line gives the results from the

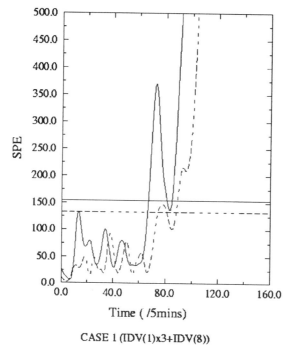

CASE 1 (IDV(1)x3+IDV(8))

Fig. 10. The SPE plot for case 1 with solid line stands for predictive monitoring and dash line for real time monitoring.

predictive monitoring strategy, and the dashed line give the results from the real time monitoring strategy which are the same as those in Fig. 8. One can see that the predictive monitoring strategy provides earlier detection of the process faults. In case 1, the SPE goes above its control limits about 10 steps (50 min) earlier than the dynamic, non-predictive result. The advances are 30 steps (150 min) in case 2 and 20 (100 min) steps in case 3. In the score plots, which are not shown here, the advantage of the predictive monitoring strategy exists in the subspace defined by the last three score variables $(t3, t4, t5)$, in which the scores from the predictive monitoring strategy go outside their limits faster than those from the dynamic, non-predictive monitoring strategy. These results show that the predictive monitoring strategy is a promising approach for dynamic process monitoring.

4. Design of a Multivariate Statistical Controller

4.1. *Description of the control problem*

One characteristic of the data collected from a multivariate continuous process is the large number of variables. Since the variables are almost never independent from one another, the true dimension of the space in which the process moves is usually very much smaller than the number of measured variables. As a result many of the measured variables move together because of a few underlying fundamental events that affect the entire process, and this leads to the latent variable approach used here. A low dimensional subspace defined by the score variables is developed using statistical data compression techniques, in this case, MPCA incorporating lagged inputs.[16] Based on past knowledge of good and bad product qualities using historical data, one can define an acceptable region of operation in the subspace. In this way, the score variables are implicitly correlated with quality variables. In the case when quality variables are not available on-line or have long time delays, the control problem for process performance can be formulated as reducing the region within which the latent variables fall. When operation is maintained inside this region, the process is functioning as it did when good product qualities were obtained. If the process deviates from this region, suitable control actions are calculated to return the process back to the good region. This is the concept, which was first presented in Piovoso *et al.*,[21] that is applied here. Earlier, a steady state statistical controller using PCA was proposed,[1] which aimed at reducing the variance of process outputs at steady state by reducing the region within which the scores fall. Here, the statistical control approach is extended to dynamic systems.

4.2. *Data representation and generation*

In our modeling approach dynamic MPCA is used to filter raw data. Then a score predictive model is developed using the dynamic MPCA scores. The data structure is the same as that shown in Fig. 4. With an appropriate order, which is the window length, a data set with this structure can capture the dynamics of a process.

The statistical controller proposed here is usually developed from an existing conventional control system, where variables such as levels, temperatures and pressures, are under control. The set points of some control loops in this control system can be chosen as the manipulated variables for the statistical controller. In some cases, prior knowledge of a process is important for manipulated variable selection. Correlation analysis between quality variables and the candidate manipulated variables can be helpful for selecting suitable manipulated variables, if one has access to an appropriate data base. In some cases, the set points for manipulative variables are fixed. The manipulated variables themselves can change but their change may not be large enough to allow their correlation to product variables to determined. One important issue is that when the statistical controller is implemented, it will change the correlation structure of the original control system. It has been pointed out[19] that the data from the original system cannot be used when a new correlation structure exists. Thus, experimental testing on the selected manipulated variables is required for the development of the predictive model. The case studies illustrate the selection of suitable manipulated variables and the experimental testing.

4.3. *Control problem formulation*

The selected score variables explain the maximum variance of X over the time horizon of each $x(t)$. Because the reference data are scaled to zero mean before building a dynamic PCA model, the center of the low dimensional subspace spanned by the score variables corresponds to the desired operation where product qualities are under perfect control, assuming process changes are only caused by random disturbances. If the disturbance structure remains the same and X correlates with product quality variables, a small region around this score point represents process operation which has a small variance of product qualities. One can set this score point as the set point of a multivariate statistical controller. If the steady state values of process variables are known, the score set point can also be obtained by projecting the steady state values onto the score space. In the case where product quality measurements are not available on-line and unmeasured random upsets occur, the control objective for process performance is defined as minimising the distance between the current score point and the score set point. The result of the proposed algorithm is a reduction in the variance of product qualities.

The major objective of the statistical controller is to compensate for the effects of random disturbances as fast as possible. However, if a new disturbance, e.g. a step, occurs, the correlation structure of the process measurements will change. It is necessary to have a criterion to judge whether a PCA model is valid for control. The SPE is used here. The control limit for the SPE can be calculated from the reference data. At each time step an SPE value is calculated for the current data window. If the SPE is above its control limit, the statistical controller should be turned off because the correlation structure of the process variables has changed. Also, the SPE should also be included in the statistical control algorithm to maintain the correlation structure among variables and their histories.

However, there are cases when a step change disturbance occurs, and as a result the means of the input variables to the PCA model change but the SPE remains under its control limit. In such a case, if the statistical controller is the only controller for quality control, the quality variables probably will have offset from their set points. One way to eliminate this offset is to close feedback loops around the quality variables, even when their measurements have long time delays. By combining the statistical and quality feedback controllers together, a large variance due to random disturbances can be shrunk and steady state offset can be eliminated. The method of integrating these two types of controllers is illustrated in the Tennessee Eastman case study below.

To carry out the optimisation in real time and to incorporate the SPE constraint easily, model predictive control (MPC) is used. The controlled variables are score variables defined by the variation from normal operation. Model predictive control (MPC) requires a dynamic predictive model for the score variables. A nonlinear autoregressive with exogenous input (NARX) model is considered in order to present the most general form of the score predictive algorithm:

$$y(t) = F(y(t-1), \ldots, y(t-n_y), u(t-1), \ldots, u(t-n_u)) + v(t)$$

where $F(\bullet)$ is a vector-valued function. $y(t)$, $u(t)$ and $v(t)$ are the score vector, the vector of manipulated variables, and noise vector, respectively, with appropriate dimensions. n_y and n_u are the time lags in the score variables and manipulated variables respectively. In order to circumvent ill-conditioning due to highly correlated data, a partial least squares (PLS) method is used for this modeling task. If the process is highly nonlinear, the NNPLS approach (Qin and McAvoy, 1992) can be used. Although the reference model may be linear, the control problem is formulated as a quadratic programming (QP) problem. The problem is stated as follows:

$$\min_{u(t)} E(t) \triangleq \sum_{k=t+1}^{t+P} \{\|\Theta e(k)\|^2 + \|\Lambda \Delta u(k=1)\|^2 + \|\Gamma \Delta y(k)\|^2\}$$

subject to:

$y(t) = PCA(x(t))$

$y'(t+1) = F[y(t), y(t-1), \dots, u(t), u(t-1), \dots]$

$y'(t+2) = F[y'(t), y(t), \dots, u(t+1), u(t), \dots]$

$\dots,$

$y'(t+P) = F[y'(t+P-1), y'(t+P-2), \dots, u(t+P-1), u(t+P-2), \dots]$

$\|x(t) - x'(t)\|^2 < \varepsilon$

where x is the vector of process measurements, y is the score vector, u is the vector of manipulated variables, and t is the current time. The error vector of the score variables is denoted as e, and d results from plant-model mismatch and unmeasured disturbances, and these variables are given by:

$$e(k) = y_{sp}(k) - y'(k) + d(t); \quad k = t+1, \dots, t+P$$
$$d(t) = y(t) - y'(t); \quad y'(t) = F[y(t-1), \dots]$$

Θ, Λ, and Γ are diagonal weighting matrices. P is the predictive horizon and M is the control horizon. There are two penalty terms, $\Gamma \Delta y$ and $\Lambda \Delta u$, included in the objective function. The penalty term on Δy can be useful when the controlled variable cannot be changed drastically, and $\Lambda \Delta u$ restricts the magnitudes of the changes of manipulated variables. Because the proposed multivariate statistical controller is built on top of an existing control system and the set points of some control loops are used as manipulated variables, it is necessary to restrict the controller from changing the original system too drastically by penalizing both Δy and Δu. A convenient way to tune the controller is to keep one penalty term constant and change the other term. To avoid additional difficulties in tuning the controller, P and M are fixed at the design stage. Guidelines on choosing P and M have been given by several authors.[2,6]

4.4. *The implementation of a multivariate statistical controller*

The procedure for building a multivariate statistical controller and its real time implementation is described below:

- The selection of suitable manipulated variables. Suitable manipulated variables, which are usually the set points of existing control loops, are selected by correlation analysis or from prior knowledge of a process.

- The development of reference data. The reference data should cover the range of the changes of dynamic patterns and the conditions which yield desired product quality outputs, and the statistical controller should be used under the same conditions. Experimental testing is needed on the selected set points used as manipulated variables in order to keep the correlation structure unchanged after the statistical controller is implemented. In the case where the statistical and quality feedback controllers are integrated, the testing should carried out and the reference data should be collected with the quality feedback loops closed.
- The data pretreatment.
- The development of a dynamic PCA model.
- The development of a score predictive model.
- The formulation of the MPC algorithm.
- The implementation of the multivariate statistical controller. At a new sampling time, the scores corresponding to the data window with previous data and the current data are calculated. The current SPE is calculated. If the SPE is above its control limit, the statistical controller is turned off and the operator is alerted. Otherwise, the MPC computation is carried out with future scores predicted from the score predictive model. Only the calculated values of manipulated variables corresponding to next sampling time are implemented on the process. The entire calculation is repeated at the next sampling instant.

The proposed multivariate statistical controller is demonstrated in the following case studies.

4.5. Examples

Two examples are given to illustrate the multivariate statistical controller. One involves a binary distillation column and the other the Tennessee Eastman process. Although the statistical control algorithm is formulated in a general NLMPC problem framework above, linear methods are used to build the PCA model and the score predictive model for these two cases. The reason is that only one operating point is considered in each case and the processes are approximately linear around the operating point considered.

4.5.1. Distillation column

This example is given to illustrate the main points of the methodology. This example involves a medium purity binary column, which separates benzene and toluene using 18 trays, with the feed on tray 9. The following assumptions are made: constant molar overflow (CMO), 100% stage efficiency, constant relative volatility, and constant column pressure. The characteristics of the column are: relative volatility = 1.5, reflux ratio = 4.29, feed composition = 0.5, top composition = 0.98, bottom composition = 0.02. A solution for distillation rating is obtained by using the Smoker equation.[26] Only one disturbance is considered here, namely random feed composition changes $\Delta x_f (\pm 10\%)$. The major control objective is to keep the top and bottom product compositions as close to their set points as possible in spite of fluctuations in the feed composition. Two available valves control the reflux flow and the vapor boilup. A typical conventional control system uses two tray temperature controllers at tray 16 and tray 4 cascaded to the top and bottom composition controllers. The sampling interval for temperatures and flow rates is every second. Both top and bottom composition measurements are assumed to have a one half hour time delay. The product composition controllers with delayed sampled inputs have large variance of product compositions for the disturbance considered. In this case study, because only a random disturbance is considered, the statistical controller does not produce offset in the compositions at steady state. A combined random plus step disturbance is considered in the next example. A multivariate statistical controller is built to replace the composition controllers.

It can be observed that when the distillation process is subject only to its natural variability (random feed composition upset), the process moves around its designed operating state, where product compositions are at their set points and no feed composition upset exists. Here, the composition measurements are assumed to be unavailable on-line. A dynamic PCA model built from tray temperature measurements is used to represent the process. The set points of two temperature controllers are used as the manipulated variables for the multivariate statistical controller. In order to build a controller, a database needs to be developed by forcing the set points of the tray temperature controllers while the normal upset (random feed composition upset) is present. The reference set contains 1000 data points with a sampling interval of 2 min. The data are rearranged into a two-dimensional array, and the length of the data window is 5 steps (10 min). Each sample is offset by 2 min in the X array. There are 996 data windows, and each window contains 5 values for each of the 18 tray temperatures used. Since data after the 1000 points used in the reference set are not available, this fact results in the number of data windows being less than 1000. A dynamic PCA model is developed from this data array, and four principal components are selected, which capture 88.63% of the variation in the reference set. The origin of the score space spanned by these principal components corresponds to the desired operating state, where product compositions are under perfect control and no feed composition upset exists. The set point of the controller is fixed at the origin of the score space.

A partial least squares (PLS) model is developed as the score predictive model, which is an ARX model. The PLS model inputs are past scores and past set points of the temperature controllers. The time lags in the score and manipulated variables are chosen to be $n_y = 3$ and $n_u = 2$ respectively, and these values produce reasonable results. The control objective function for this case study includes only one penalty term, $\Lambda \Delta u$, where Λ equals 150 times an identity matrix with an appropriate order. There are no explicit upper and lower bounds for u, Δu, y, and Δy. Other parameters are: $P = 4$ and $M = 3$. The control results from the multivariate statistical controller and the PI composition controllers are shown in Fig. 11. Plot (a) gives

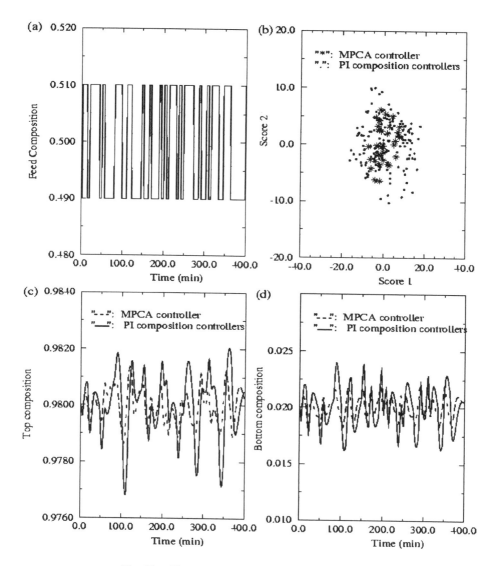

Fig. 11.   The control results on a distillation tower.

the feed composition and Plot (b) is the comparison in the score space between results from the multivariate statistical controller and those from the PI composition controllers with delayed composition measurements. One can see that the scores from the multivariate statistical controller cover a smaller region around the origin than those from the PI composition controllers. Plots (c) and (d) give the control results for top and bottom compositions respectively. The solid lines correspond to the results from the PI control system, and the dash lines are the results from the multivariate statistical controller. The figure shows that the multivariate statistical controller is better than the PI composition controllers with delayed sampled inputs. The multivariate statistical controller achieves a smaller variance of product compositions by detecting and compensating for the effect of an unmeasured feed composition disturbance faster than the PI controllers. Further study shows that the multivariate statistical controller gives results that are comparable to a PI control system that has no analyser delay. This result demonstrates that the multivariate statistical controller is a promising tool for improving the performance of a plant control system without expensive on-line analysers.

4.5.2. *The Tennessee Eastman Process*

Again, the Tennessee Eastman Process is used to illustrate the design of a dynamic statistical controller. Here, a 50/50 G/H product ratio under the base control system[25] is considered. Most measurements, such as flows, temperatures, and pressures, are available every second. However, the sampling frequency of the product analysis is 0.25 h and as a result product composition measurements have a 0.25 h dead time. The base control system uses delayed G and H product composition measurements for G/H product ratio control. The variance of the G/H product ratio is large when random disturbances occur in the C feed (IDV(8)).

This case study is aimed at reducing the variance of the G/H product ratio caused by the unmeasured random disturbance. A total of 11 input variables are selected from the Tennessee Eastman process for building the PCA model. These 11 input variables include the variables not used in the conventional control system, as well as controlled variables of inner cascade loops, and they are listed in Table 2.

Selection of inputs for the predictive score controller is an important consideration. In an earlier study (McAvoy *et al.* (1996)) an inferential controller was developed for the Tennessee Eastman plant based on its steady state model. Most of the measurements used in this earlier study are used here. The two measurements that are not used are the compressor recycle valve position (lack of sensitivity), and the product flow (slow response). If such prior knowledge were not available, one would have to rely on plant experience in choosing input variables, or carry out

Table 2.   Input variables of the MPCA model.

Process variables	
XMEAS(1)	A Feed (stream 1)
XMEAS(3)	E Feed (stream 3)
XMEAS(6)	Reactor Feed Rate (stream 6)
XMEAS(10)	Purge Rate (stream 9)
XMEAS(11)	Product Sep Temp
XMEAS(13)	Prod Sep Pressure
XMEAS(14)	Prod Sep Underflow (stream 10)
XMEAS(16)	Stripper Pressure
XMEAS(19)	Stripper Steam Flow
XMEAS(21)	Reactor Cooling Water Outlet Temp
XMEAS(22)	Separator Cooling Water Outlet Temp

a correlation analysis. The selection of manipulated variables for the multivariate statistical controller is mainly based on prior knowledge of the Tennessee Eastman process. From the material balance of the process, one can see that the D/E ratio in the input to the plant determines the G/H product ratio.[25] Since the E feed flow is used to control the reactor level, the D feed flow and the reactor level setpoint are good candidates for the manipulated variables for the multivariate statistical controller.

First, only the D feed flow is used as a manipulated variable for the statistical controller. Normal operation is defined as the control response to random A, B, C composition upsets in feed stream 4 (disturbance IDV(8)). When building a reference data set, the G/H ratio control loop is kept the same as that in the base control system[25] in order to eliminate the steady state offset of the G/H ratio. At the same time, a pseudo-random multi-step sequence (PRMS) signal is added to the D feed set point. The maximum magnitude of the PRMS is 5% of the steady state value of the D feed flow. The reference set contains 1000 samples from normal operation with a sampling interval of 5 min. The time constants of the control loops range from several minutes to more than 10 h and a window length of 50 min (10 steps) is selected. Each sample is offset by 5 min in the X array. There are 990 data windows, and each window contains 10 values for each of the 11 input variables used. A dynamic PCA model is developed from this data array. Five principal components are selected, which capture 61.27% of the variation in the reference set. The origin of the score space spanned by these principal components corresponds to the initial steady state condition. The origin is used as the set point of the statistical controller.

The score predictive model is an ARX model built using PLS. The time lags in the score variables and D feed flow set point are chosen to be $n_y = 3$ and $n_u = 10$ respectively. Both penalty terms, $\Gamma \Delta y$ and $\Lambda \Delta u$, are used in the control object function. The weighting matrices are $\Gamma = 0.5 * I$ and $\Lambda = 10 * I$, where I stands for identity matrices with appropriate orders. It is stated in the Tennessee Eastman problem[5] that the D feed should not have significant frequency content in the range from 8 to $16 \, h^{-1}$. The penalty on $\Lambda \Delta u$, namely the change of D feed set point, is chosen to satisfy this constraint. The upper and lower bounds for y are set as the 95% confidence limits of the score variables, which are determined from the reference data by using the methodology presented by Nomikos and MacGregor.[20] The upper and lower bounds for u are set based on the maximum variation of u during the normal operation defined above. The maximum variation of the D feed flow is $\pm 7.0\%$ of its steady state value. Other parameters are chosen as: $P = 5$ and $M = 5$. The output of the statistical controller is designed to be the change of the D feed set point and it is added to the D feed set point calculated from the PI composition controller. In this way, the outputs from the statistical and composition feedback controllers are combined together and the effect of the statistical controller is also restricted to avoid changing the D flow too drastically. The SPE hard constraint is not reached in this case and the statistical controller is active all the time.

The control results for the random feed composition fluctuations (IDV(8)) are shown in Fig. 12. The outputs (changes to the D feed flow set point) from the statistical controller and the PI composition controller are shown in Fig. 13. The solid lines correspond to the results from the base control system with the PI product composition controller only, and the dash lines result from the statistical controller. One can see that the statistical controller achieves a smaller variance of product compositions by detecting and compensating for the effect of unmeasured feed composition disturbances faster than the feedback PI controller. Figure 13 shows that the manipulated D flow from the statistical controller responds quicker than it does in the base control system.

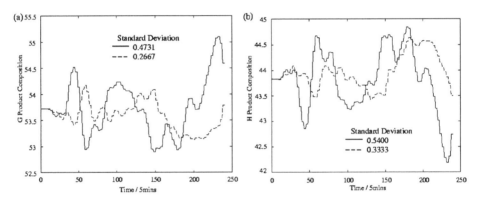

Fig. 12. The control results from the multivariate statistical controller on D feed flow and the base control system. (a) G product composition; (b) H product composition. (Solid line: Base control system; dash line: Multivariate statistical controller.)

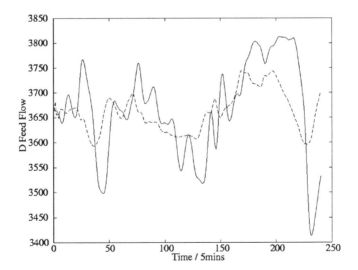

Fig. 13. The controller outputs from the multivariate statistical controller on D feed flow and the base control system. (Solid line: Base control system; dash line: Multivariate statistical controller.)

The other test concerns the steady state offset of the G/H product ratio. In this test, the disturbance IDV(8) is combined with a large step change in A/C composition ratio in feed stream 4, whose magnitude is 50% larger than that of the original disturbance IDV(1) (disturbances IDV(1)x1.5 + IDV(8)). The large magnitude of the step change is aimed to clearly show the offset problem that can arise. However, the magnitude is not too large to cause the SPE to move outside its control limit. A statistical controller is built and implemented with the G/H ratio feedback loop open and it has the D feed set point as its manipulated variable. The control results for the G/H ratio from the base control system and this statistical controller are shown in Fig. 14. One can see that the G and H compositions have a steady state offset when only the statistical controller is used and the step disturbance occurs. Integrating the statistical and composition feedback controllers together can solve

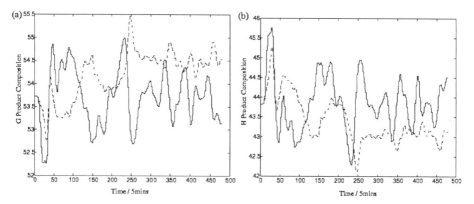

Fig. 14.   The control results from base control system and the multivariate statistical controller on D feed flow with G/H ratio PI loop open. (a) G product composition; (b) H product composition. (Solid line: Base control system; dash line: Multivariate statistical controller.)

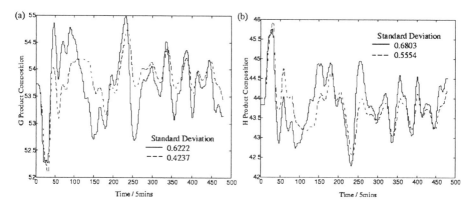

Fig. 15.   The control results from base control system and the multivariate statistical controller on D feed flow with G/H ratio PI loop closed. (a) G product composition; (b) H product composition. (Solid line: Base control system; dash line: Multivariate statistical controller.)

this problem. The statistical controller from the first case is used again for this case. Figure 15 shows the control results from the base control system and this statistical controller. From the figure, one can see that the combination of the statistical and composition feedback controllers has the ability to track the G/H set point and at the same time the variance of the product compositions is reduced.

5. Conclusions

The approaches for dynamic process monitoring using Multi-way Principal Component Analysis (MPCA) and designing a dynamic statistical controller have been presented. In order to consider the auto correlation of process variables, dynamic data are used in the MPCA model. The predictive monitoring strategy has a timing advantage over traditional monitoring approaches based on PCA. The proposed monitoring strategy is demonstrated on the Tennessee Eastman process. The results show that the proposed approach is able to provide more rapid detection of operating problems than previously published approaches. The predictive monitoring approach presented holds promise as an effective means of using the readily available data to solve the problem of monitoring dynamic processes. The statistical controller is based on the definition of a new control set point within the subspace developed from the dynamic MPCA model. Such a controller belongs to the class of model based controllers and it can be designed under the nonlinear model predictive control (NLMPC) framework. The multivariate statistical controller is demonstrated on a binary distillation column and the Tennessee Eastman process. The results show that the controller is effective in reducing the variance of product quality variables caused by the same disturbances with the same magnitude as occurred during the data collection. The multivariate statistical controller utilises set points in a conventional control system to improve control performance. The multivariate statistical controller involves an approach to use normal operating data coupled with limited plant testing, and it can be added on top of an existing conventional control system to improve process performance.

References

1. G. Chen and T. J. McAvoy, Process control utilizing data based multivariate statistical models, *The Canadian Journal of Chemical Engineering*, 1996.
2. C. Chien, I. Lung and P. S. Fruehauf, Consider IMC tuning to improve controller performance, *Chemical Engineering Progress* **86**, 10 (1990) 33.
3. R. Cutler and B. Ramaker, Dynamic matrix control — A computer control algorithm, *AIChE Annual Meeting*, Houston, TX, 1979.
4. D. Dong and T. J. McAvoy, Nonlinear principal component analysis based on principal curves and neural networks, *Comput. Chem. Eng.* **30**, 2 (1996) 65.
5. J. J. Downs and E. F. Vogel, A plant-wide industrial process control problem, *Comput. Chem. Eng.* **17**, 3 (1993) 245–255.
6. C. E. Garcia and M. Morari, Internal model control 1., A unifying review and some new results, *Ind. Eng. Chem. Process Des. Dev.* 21 (1982) 308–323.

7. C. E. Garcia, D. M. Prett and M. Morari, Model predictive control: Theory and practice — A Survey, *Automatica* **25**, 3 (1989) 335–348.

8. R. Geladi, Analysis of multi-way (multi-mode) data, *Chemometrics and Intelligent Lab. System* **7** (1989) 11–30.

9. P. Geladi and B. Kowalski, Partial least squares regression: A tutorial, *Analytica Chimica Acta* 185 (1986) 1–17.

10. T. Hastie and W. Stuetzle, Principal curves, *J. Am. Statistical Assoc.* **84**, 406 (1989) 502–516.

11. J. E. Jackson, *A User's Guide to Principal Component Analysis* (John Wiley, New York, 1991).

12. I. T. Jolliffe, *Principal Components Analysis* (Springer Verlag, New York, 1986).

13. K. A. Kosanovich, M. J. Piovoso, K. S. Dahl, J. F. MacGregor and P. Nomikos, Multi-way PCA applied to an industrial batch process, *Proc. ACC*, 1994, 1294–1298.

14. J. V. Kresta, J. F. MacGregor and T. E. Marlin, Multivariate statistical monitoring of process operating performance, *The Candian Journal of Chemical Engineering* **69**, 2 (1991) 35–47.

15. J. Kresta, T. E. Marlin and J. F. MacGregor, Choosing inferential variables using projection to latent structure (PLS) with application to multicomponent distillation, *Proc. AICHE Annual Meeting*, Chicago, IL, 1990.

16. W. Ku, R. Store and C. Georgakis, Disturbance detection and isolation by dynamic principal component, *Chemometrics and Intelligent Lab. System* **30** (1995) 179–196.

17. L. Leontaritis and S. Billings, Input-output parametric models for non-linear systems: Part 1, Deterministic non-linear systems; Part II, Stochastic non-linear systems, *Int. J. Control* **41** (1985) 303–344.

18. J. F. MacGregor, Multivariate statistical methods for monitoring large data sets from chemical processes, *Annual AICHE Meeting*, 1989.

19. J. F. MacGregor, T. E. Marlin and J. V. Kresta, Some comments on neural networks and other empirical modeling methods, CPC-IV, Austin and New York, 1991, 665–672.

20. R. Nomikos and J. F. MacGregor, Monitoring of batch processes using multi-way PCA, *AICHE J.* **40** (1994) 1361–1375.

21. M. J. Piovoso, K. A. Kosanovich and P. K. Pearson, Monitoring process performance in real time, *Proc. ACC*, 1991.

22. M. J. Piovoso, K. A. Kosanovich and J. P. Yuk, Process data chemometrica, *IEEE Trans. Instrum. Measure.* **41**, 2 (1992) 262–268.

23. M. J. Piovoso and K. A. Kosanovich, Applications of multivariate statistical methods to process monitoring and controller design, *Int. J. Control* **59**, 3 (1994) 743–765.

24. W. A. Shewhart, *Economic Control of Quality Manufactured Product* (Van Nostrand, Princeton, New Jersey, 1931).

25. L. Stahle, Aspects of the analysis of three-way data, *Chemometrics and Intelligent Lab. System* **7** (1989) 95–100.

26. L. T. Tolllver and R. C. Waggoner, Approximate solutions for distillation rating and operating problems using the smoker equations, *Ind. Eng. Chem. Fundam.* **21** (1982) 422–427.

27. S. Wold, P. Geladi, K. Esbensen and J. Ohman, Multi-way principal components and PLS analysis, *J. Chemomet.* **1** (1987) 41–56.

28. N. Ye and T. J. McAvoy, An improved base control for the Tennessee Eastman problem, *Proc. of ACC* (1995) 240–245.

INDEX